Lecture Notes in Computer Science

Edited by G. Goos and J. Hartmanis

445

A.J.M. van Gasteren

On the Shape of Mathematical Arguments

Foreword by Edsger W. Dijkstra

Springer-Verlag
Berlin Heidelberg New York London
Paris Tokyo Hong Kong Barcelona

Author

Antonetta J. M. van Gasteren
University of Groningen, The Netherlands
and
University of Utrecht, The Netherlands

Correspondence address:
Edelweislaan 20
5582 BW Waalre, The Netherlands

CR Subject Classification (1987): F.3.1, D.2.4

ISBN 3-540-52849-0 Springer-Verlag Berlin Heidelberg New York
ISBN 0-387-52849-0 Springer-Verlag New York Berlin Heidelberg

© Springer-Verlag Berlin Heidelberg 1990
Printed in Germany

Printing and binding: Druckhaus Beltz, Hemsbach/Bergstr.
2145/3140-543210 – Printed on acid-free paper

Foreword

While current curricula extensively teach existing mathematics, they pay scant attention to the doing of mathematics, i.e., to the question of how to design and to present solutions. If any attention to these issues is paid at all, they are treated separately: design of solutions, i.e., "problem solving" or "mathematical invention", is viewed as a psychological issue, as a matter of mathematical intuition, while presentation is viewed as a matter of personal style or as an issue of education. Most mathematicians consider psychology and pedagogy as sciences too soft to be respectable, and consequently the subject of how to do mathematics has almost been tabooed.

The great merit of A.J.M. van Gasteren's work is to have broken this taboo. She has done so, firstly, by recognizing that, in the case of mathematical arguments, the traditional separation of content and form is untenable, and, secondly, by showing how, via notation and formula manipulation, design and presentation emerge as two sides of the same coin. She has broken the taboo because, in this united setting, the issues involved are purely technical: for instance, the question of whether a certain notational convention —of equal relevance for the derivation as for the presentation of a solution!— is geared to our manipulative needs is a technical question that has nothing to do with intuition or taste.

That formalization should and can aid the mathematician rather than add to his burden is well-known. This book reveals a wider vision, viz., that the more formal the solutions we are heading for, the better we shall be able to teach the art and science of their design. In doing so, this work represents a crucial first step towards mathematical methodology in the truest sense of the word, and for that reason I greatly welcome its publication in Springer's Lecture Notes Series: the more people enjoy this book and are inspired by it, the better.

Edsger W. Dijkstra

III

Preface

For me, exploring the presentation and design of mathematical proofs has been a very exciting activity. To see how virtually every proof and program can be made much clearer and crisper, and to develop techniques and skills for doing so, is such a rewarding experience that my hope is to share it with the readers of this monograph. To appreciate the text, the reader needs some maturity in mathematics, but no specialized mathematical knowledge is required to follow the example arguments in the first part.

Of all those who made my explorations and the writing of this text possible, I want to mention first and foremost Edsger W. Dijkstra and W.H.J. Feijen, who educated me and who shared and guided my interests. The Eindhoven Tuesday Afternoon Club, in varying formations, has always been an indispensable forum for discussion and critique. In their capacity as members of my dissertation committee, R.C. Backhouse, P.L. Cijsouw, and F.E.J. Kruseman Aretz carefully studied, and commented on, this text.

For many years, BP International Limited provided financial support for my research under their much appreciated Venture Research Scheme; especially D.W. Braben, head of BP's Venture Research Unit, has been an ever interested and stimulating ally. V.H. Backhouse and D.B.M. Klaassen took care of the typesetting and typing of this text, accepting and implementing all my well-considered but non-standard typographic wishes. H. Paas of the University of Groningen provided them with technical assistance. M.C. Dijkstra-Debets and D.B.M. Klaassen, finally, deserve mention for their moral and other support.

Contents

0   Introduction   0

Part 0   8

Summaries of the expositional essays   10

1   A termination argument   14

2   A problem on bichrome 6-graphs   17

3   Proving the existence of the Euler line   21

4   In adherence to symmetry   25

5   On a proof by Arbib, Kfoury, and Moll   29

6   Not about open and closed sets   34

7   A monotonicity argument   41

8   On the inverse of a function   45

9   A calculational proof of Helly's theorem on
    convex figures   47

10 The formal derivation of a proof of the
    invariance theorem   57

11 Proving theorems with Euclid's algorithm   64

12 On the design of an in-situ permutation algorithm   69

13 Shiloach's algorithm   80

Part 1   88

14 Clarity of exposition   90

14.0 The division of labour between reader and writer 91

14.1 On arrangement 107

14.2 On disentanglement 116

15 On naming   122

15.0 Names in a linguistic context 122

15.1 Names in a formal context 126

16 On the use of formalism   136

16.0 Manipulation without interpretation 138

16.1 On convenience of manipulation 144

16.2 Avoiding formal laboriousness 155

17 Epilogue   166

18 Proof rules for guarded-command programs   171

19 Notational conventions   174

References   178

0 Introduction

This is a study about proofs; it is not about theorems or theories, but about mathematical arguments, proofs of correctness for programs included. The original incentive to investigate proofs in their own right was simply a matter of necessity; later the challenge of exploring a topic that has received relatively little explicit attention became a second impetus for the investigations.

Let me briefly discuss the necessity. In the late seventies it had become possible to derive a program and its correctness proof hand-in-hand, and if the proof was sufficiently detailed the outcome was a trustworthy design. That constraint of sufficient detail, however, was something of a problem, because what mathematicians usually consider sufficient detail did not suffice in the context of program design, and for all but the simplest problems the requirement, if met at all, tended to lead to proofs that were long and verbose, or complicated and laborious when given formally.

Our conclusion was that computing scientists would have to learn to make their proofs more effective, if their methods were to be applicable to more ambitious problems as well. At the time, however, it was far from obvious what the characteristics of effectiveness could be (although one thing *was* clear: it should be a combination of completeness and brevity of argument), and even less obvious how they might be effectuated.

That was, in short, the initial incentive for investigating proofs in their own right. All by itself the topic is so rich that the present study can only be considered the beginning of a much larger exploration. The explorations reported here have been aimed at variety rather than at concentration on a few special topics, so that now we have gathered a lot of themes each deserving more extensive exploration all by itself.

To avoid confusion and misunderstanding, it seems appropriate to delimit the scope of this study.

•   Firstly, the major stress is on presentation rather than on design. Although in the long run the latter is the more important and interesting topic, there was, I believe, a very good reason for postponing its investigation: it is hard to imagine how one can become articulate about methods of finding proofs without knowing what kind of proofs one would like to find. Besides that, the relative scarcity of literature on the topic suggested that a study of expositional issues could in itself be a valuable endeavour.

Thus, this study is more concerned with form than with content —whence its title— , although the distinction between the two turned out to be less sharp than we had anticipated. We found, for instance, that the choice of nomenclature, usually considered to belong to the realm of presentation, could greatly influence the structure of the argument.

By this and other experiences, heuristic considerations ultimately started to play a somewhat larger rôle; wherever appropriate they will be elaborated upon. As a rule, these heuristic considerations do not address psychological questions like: "How do people find solutions"; they address more technical questions like: "How can proof design be guided by syntactic analysis of the demonstrandum". That perhaps not many mathematicians to date arrive at their proofs by means of such a technique is only of secondary importance. In order to improve upon the status quo one must be willing to deviate from it, in matters of heuristics as well as in matters of exposition or notation. (In our explorations we ignore the practical problems of the day that may arise from such a deviation, like the difficulty of unfamiliar conventions or the constraints of editorial policy. That does not mean, however, that we consider such problems negligible.)

• Because of the goal of combining completeness and brevity of argument, the recourse to formalism presented itself from the very beginning, and, indeed, the use of formalism plays a predominant rôle in this study. This holds particularly for the use of predicate calculus.

That does not mean, however, that this study is concerned with foundations. The properties of interest here, viz. properties that make a formalism convenient for use, are not the same as those properties that make it convenient for study: while for the logician the existence of proofs is a major point, the user of a formalism is more interested in their efficiency; while redundance of the rules is inconvenient for the study of a formalism, that doesn't necessarily hold for its use; while for the logician the distinction between axioms, theorems, and "metatheorems" is relevant, it is not for the user, whose main concern is their validity.

Likewise, while in this study the use of equivalence is stressed more than the use of implication, and implications $P \Rightarrow Q$ are sometimes replaced by the equivalent $P \wedge Q \equiv P$, that does not automatically imply preference for a calculus based on equivalence rather than on implication.

• This study deals with "human" theorem proving, and not with mechanical verification or mechanical theorem proving. I consider it important to mention this choice explicitly, because superficially some of the interests shown in this study might suggest the opposite. For instance we share, with those involved in mechanical verification and proof design, the interest in manipulation —of formulae— without interpretation and in proof design guided by syntactic analysis of the demonstrandum. Such similarities indicate that efforts at mechanization might contribute to "human" theorem proving more than is already the case.

(On the other hand the differences are large. Although efficiency is a common concern, we first of all aim at efficiency of the result, viz. of the proof, and hence at short formulae and short derivations, and at the avoidance of repetitiousness in formulae and argument; for the designers of mechanical systems the efficiency of the process of constructing a proof is the major worry. Another difference is the usability of symbolic rewrite rules, e.g. $A \equiv B$. In mechanical systems they are not so popular; they need special treatment because they can lead to nonterminating derivations: replacements of one side by the other can be undone again; the present study, however, extensively exploits such rewrite rules at advantage.)

• Although the explorations reported here have been inspired by computing's needs and challenges, mathematical proofs in general were an object of investigation just as well as correctness proofs of programs. They did so because of my personal interest and because I expected both fields to profit from each other. In addition, their inclusion provided a wealth of extra source material.

Although the work was not confined to programming alone, earlier experiences in the development of programming methodology and needs of the field did have an influence on how the explorations were conducted, on what was done first and what was considered most important. Some of these influences are listed below.

Firstly, the decision to postpone heuristics was inspired by a similar decision taken in the early days of the development of programming methodology: the latter began to make progress only after it had been realized that not every program is worth proving. When it had become clearer what a "nice" program might look like, drastic simplifications often proved to be possible. Programming similarly inspired the decision to experiment extensively with "smaller" problems first, rather than tackling a necessarily small number of "large" problems with the danger of having to deal with too many details specific to the particular problem.

Furthermore, the decision to explore the use of formalism extensively and, more generally, the stress on methodological issues were an immediate consequence of what we felt to be needed in computing and potentially useful in general. The greater concern with methodology is the consequence of the fact that computing science is one of the less knowledge-oriented branches of applied mathematics.

For instance, for the computing scientist, the technique of "reducing" a problem to an already solved one is not nearly as obviously appropriate as it is for the mathematician that wishes to establish the validity of some hypothesis: for a programming problem not only the existence of a solution but also the solution's efficiency —in terms of computation time and space— is vital, and reduction of a problem to an already solved one does not necessarily give the most efficient program.

In addition to this, computers really deserve the qualification "general purpose", which means that the computing scientist is regularly confronted with problems for which the relevant concepts, notations, and theory have not been developed yet. That was another reason for stressing methodological concerns.

So much for some of the ways in which computing has had an influence on the explorations.

End • .

The scope of the investigations having been delimited, the next point perhaps is what one can expect as the result of such explorations. After all, many hold the opinion that mathematical and expositional style are purely (or at best largely) a matter of personal taste. Admittedly, there is no such thing as the "best proof" or the rule of thumb that always works, but what I hope to show is the existence of a variety of technical criteria by which one argument can be objectively less streamlined than another, and a number of expositional alternatives that are, for unclear reasons, neglected. It has turned out that a lot can be said about mathematical arguments in general that is independent of the particular area of mathematics an argument comes from. Among the topics explored are, for instance, proofs by cases, exploitation of symmetry, the problem of what to name, the exploitation of equivalence, proofs by calculation and their influence on the choice of notations, the degree of detail of proofs, and linearization of proofs.

*   *

*

One of the problems to be solved with a methodological study like this is how to sail between the Scylla and Charibdis of vagueness by too much generality on the one hand and explanation-by-example only on the other. The solution chosen here is a study consisting of two parts, viz. a series of "expositional essays" and a number of more general chapters putting the example arguments into perspective.

Each of the essays deals with one problem —a theorem to be proved or a program to be designed—. The problems themselves are of minor importance; they have been chosen for what can be illustrated by their solutions. For that purpose they have been chosen small enough to avoid raising too many issues at a time, and sufficiently diverse to show a variety of characteristics.

With one or two exceptions, each essay contains a "model" solution I think beautiful enough for inclusion and an alternative argument taken from the literature, with which the first is compared and contrasted. In one case only an argument from the literature is discussed, and for some of the programming problems the contrasting argument is absent because in the literature a correctness proof has hardly been given. In most cases, the discussion of the argument given includes remarks about the design of the proof.

I do not claim that the model arguments are the best possible, because I do not believe that such a thing exists, nor do I claim that the contrasting arguments are the best ones to be found in the literature. The latter have been chosen from the writings of traditionally reputable authors, not because I wanted to dispute their mathematical qualities, nor because the symptom of ineffectiveness discussed occurs more often in their writings than in others', nor because I think that in their cultural or historical context they could have done "better", but primarily to show that the phenomena discussed do not just occur in some obscure writings only.

So much for the expositional essays. As indicated by the chapter titles, the other part deals in a more general setting with naming, clarity of exposition, and notation and the use of formalism.

The two parts can be read independently, in either order; in fact each chapter has been written to be as self-contained as possible. References from one part into the other do occur, but they have been phrased in such a way that prior reading of the passage referred to is not strictly necessary.

Notwithstanding their independence, however, the reading of either part will probably be more profitable with some knowledge of what is in the other (it may, for instance, be instructive to read the passage on the proof format in Chapter 16 before reading the more formal expositional essays, and, conversely, to read some of the expositional essays in which naming is an issue before reading the chapter devoted to that topic). To assist the reader in choosing an order that suits him best, the series of expositional essays starts with a short description of the main points of each essay. For the sake of convenience a list of notational and other conventions and a summary of proof rules for programs in guarded command notation have been included in this book.

Part 0

Part 0.

Summaries of the expositional essays

1 A termination argument

The point of this little essay is to show in a nutshell how exploitation of symmetries —in this case between zeroes and ones— does more than reducing the length of an argument by a factor of 2 : the exploitation strongly invites the "invention" of the concept in which the argument is most readily expressed. The essay is an exercise in not naming what can be left anonymous.

2 A problem on bichrome 6-graphs

This chapter's main purpose is to show the streamlining of a combinatorial argument full of nested case analyses. The decision to maintain all symmetries is the major means to that end: the consequential avoidance of nomenclature strongly invites the use of a counting argument rather than a combinatorial one, and, like in "A termination argument", the "invention" of a concept in terms of which the argument is most smoothly formulated.

3 Proving the existence of the Euler line

This chapter is concerned with some consequences of introducing nomenclature: repetitiousness, caused by the destruction of symmetry that is inherent to giving different things different names, and lack of disentanglement, caused by the availability of avoidable nomenclature. A second point the chapter wants to illustrate —and remedy— is how the use of pictures has the danger of strongly inviting (i) the introduction of too much nomenclature, and (ii) implicitness about the justification of the steps of the argument.

4 In adherence to symmetry

This chapter is another illustration of the complications engendered by the introduction of nomenclature, here emerging in the form of over-specificity and loss of symmetry. It also discusses the choice between recursion and complete unfolding.

5 On a proof by Arbib, Kfoury, and Moll

This chapter discusses an extreme example of the harm done by the introduction of nomenclature that forces the making of avoidable distinctions, in particular the introduction of subscripted variables. In addition it illustrates some consequences of neglecting equivalence as a connective in its own right.

6 Not about open and closed sets

This chapter is primarily included as an example of orderly and explicit proof development guided by the shape of the formulae rather than by their interpretation. In passing it illustrates the usefulness of the equivalence in massaging proof obligations. In revealing the structure of our argument clearly and in justifying in a concise way why each step is taken, the use of formalism is essential.

7 A monotonicity argument

The belief that equivalence is always most appropriately proved by showing mutual implication has undoubtedly been strengthened by the way in which proofs in Euclidean geometry are conducted. The purpose of this chapter is to show that some of that "geometrical evidence" is not compelling at all.

8 On the inverse of a function

This very small essay tracks down the origin of an asymmetry in the usual treatments of the notion of the inverse of a function, and does away with that asymmetry. It is another exercise in maintaining equivalence.

9 A calculational proof of Helly's theorem on convex figures

The proof in this chapter is included firstly to show the calculational style in action, this time in a geometrical problem, and, secondly, to illustrate the carefully phased exploitation of data that is enabled by the introduction of nomenclature.

10 The formal derivation of a proof of the invariance theorem

The construction of the formal proof in this chapter illustrates to what extent the shape of formulae rather than their interpretation can inspire and assist the design of a proof.

11 Proving theorems with Euclid's algorithm

Algorithms can be used to prove theorems. This chapter illustrates how the notion of invariance can assist in proving equivalences directly instead of by mutual implication.

12 On the design of an in-situ permutation algorithm

It is shown how the availability of an adequate notation, for permutation-valued expressions in this case, can be essential for the presentation of an algorithm and the design decisions leading to it.   The choice of the notation was guided by constraints of manipulability, constraints that were met primarily by being frugal in the use of nomenclature (of subscripted variables in particular).

13 An exercise in presenting programs

This chapter's purpose is to show how the use of an adequate formalism, predicate calculus in this case, enables us to present an algorithm clearly, concisely, and in all relevant detail, in a way that reveals all the ingenuities of the design.

1 A termination argument

The point of this little essay is to show in a nutshell how exploitation of symmetries —in this case between zeroes and ones— does more than reducing the length of an argument by a factor of 2 : the exploitation strongly invites the "invention" of the concept in which the argument is most readily expressed. The essay is an exercise in not naming what can be left anonymous.

We are requested to provide an argument for the termination of the following game: a finite bit string (i.e. a string of zeroes and ones) is repeatedly transformed by replacing

a pattern 00 by 01 , or
a pattern 11 by 10 , wherever in the string and
as long as such transformations
are possible.

The argument will consist in the construction of a variant function, i.e. a function that decreases at each transformation and is bounded from below.

Since the pair of transformations is invariant under an interchange of 0 and 1 , only equality and difference of bits matter. Exploiting this observation, we record the succession of neighbour equalities and differences in the bit string as a string of y's and x's , with

15

y standing for a pair of equal neighbour bits, and
x standing for a pair of different neighbour bits

(which given the first bit precisely determines the bit string).

In this terminology, a transformation changes a y in the "code string" into an x, while leaving all elements to the left of that y unchanged. Thus the code string decreases lexically at each transformation. Since it furthermore is lexically bounded from below —by the string of appropriate length consisting of x's only— the game terminates.

(The shape of the bit string upon termination follows from the observation that the leftmost bit of the bit string does not change in the game and that upon termination the code string consists of x's only.)

*   *

*

The introduction of the code string effectively exploits the symmetry between 0 and 1, since it hides the individual bits completely. Thus we can discuss the effects of a transformation without being tempted to use case analysis.

More importantly, however, the introduction of the code string allowed us to use lexical ordering as "canned induction": our argument boils down to proving that the game terminates for each code string by observing that (i) the game terminates for the lexically smallest code string and (ii) if the game terminates for all code strings lexically smaller than a code string s, it also terminates for s, since any single transformation changes s into a lexically smaller code string. This proof by induction on lexically ordered strings is valid since lexical ordering is well-founded.

In a way the argument presented is as efficient as possible: we only had to consider one change of one symbol, viz. of a y into an x; not even the reverse change of an x into a y was relevant. One can certainly imagine other proofs, such as a proof by induction on the length of the bit string or a proof —by induction on the number of preceding bits— that each bit is changed only a finite number of times; it is hard to imagine, however, how such proofs could be more efficient.

Finally we add that, as usual, we have consciously tried to introduce only a modest amount of nomenclature. We named neither the lengths nor the individual elements of bit string and code string and learned that, indeed, no such names were needed.

2 A problem on bichrome 6-graphs

This chapter's main purpose is to show the streamlining of a combinatorial argument full of nested case analyses. The decision to maintain all symmetries is the major means to that end: the consequential avoidance of nomenclature strongly invites the use of a counting argument rather than a combinatorial one, and, like in "A termination argument", the "invention" of a concept in terms of which the argument is most smoothly formulated.

We present and discuss two expositions for the following problem. Consider a complete graph on 6 nodes, each edge of which is either red or blue; demonstrate that such a coloured graph contains at least 2 monochrome triangles. (Three nodes form a "monochrome triangle" if the three edges connecting them are of the same colour.)

Exposition0 . This exposition first establishes the existence of 1 monochrome triangle as follows. Of the 5 edges meeting at some node X, at least 3 have the same colour, say red. Calling their other end points P, Q, and R respectively, we have: triangle PQR is monochrome or it contains at least 1 red edge, PQ say. In the latter case triangle PQX is all red.

To establish the existence of a second monochrome triangle we assume that the existence of a, say, all-red triangle has been established. We mark each of its nodes "A" and each of the remaining nodes "B" .

17

Our first dichotomy is: triangle BBB is monochrome or it is not. In the latter case BBB has at least 1 red edge and at least 1 blue edge; also, any second monochrome triangle is of the form AAB or BBA.

Case BBB not monochrome is subdivided into two subcases: there is a monochrome triangle AAB —i.e. an all-red AAB, since AAA is all red— or there is not. In the latter case we hence have that at each B less than 2 red BA-edges meet; hence at each B at least 2 of the 3 BA-edges are blue. From these and a blue BB-edge, the existence of which we have not exploited yet, we find an all-blue BBA-triangle: of the at least $2 + 2$ blue BA-edges meeting at the endpoints of a blue BB-edge, 2 meet at the same A (on account of the pigeon-hole principle), thus yielding a blue BBA.

End Exposition0 .

*

The above proof, though not long, yet sufficiently detailed, is unattractive, because of its case analysis and its destroying all sorts of symmetry.

The trouble already starts with "at some node X", which by naming one node destroys the symmetry among the nodes. The next harm is done by the introduction of the three distinct names P, Q, and R: the subsequent "PQ say" shows how the naming inappropriately breaks the symmetry. (The use of subscripted names would not have been any better.)

Later the more symmetric nomenclature AAA/BBB is used, which somewhat smoothes the presentation of the second part of the argument, but still we have the A's versus the B's.

By distinguishing a first and a second monochrome triangle we had to distinguish three cases for the second, viz. whether it shares 0, 1, or 2 nodes with the first triangle.

The symmetry we lost almost from the start is the symmetry between the colours. All these distinctions render a generalization of the exposition to graphs with more nodes very unattractive if not impossible.

$$* \qquad *$$

Exposition1 is based on two decisions: to maintain the symmetry between the colours and among the nodes, even to the extent that we shall try to leave them anonymous.

Exposition1 . We head for a counting argument to establish a lower bound on the number of monochrome triangles, because such arguments are more likely to maintain symmetry. To that purpose we wish to characterize monochrome triangles, or bichrome ones —whichever is simpler— . We have this choice because in the complete 6-graph the total number of triangles is fixed, viz. 20. Hence the difference of 20 and an upper bound on the number of bichrome triangles is a lower bound on the number of monochrome triangles.

To investigate which is easier to characterize we note that for a monochrome triangle we need 3 edges of equal colour; for a bichrome one, however, 2 differently coloured edges meeting at a node suffice. The latter is the simpler characterization, because $2 < 3$ and no colour specification is needed. Therefore, we give the concept a name and investigate its properties.

A bichrome V is a pair of differently coloured edges meeting at a node.

Bichrome V's and bichrome triangles satisfy (i) each bichrome triangle contains two bichrome V's (ii) in a complete graph, each bichrome V is contained in exactly one bichrome triangle. Consequently, the number of bichrome triangles is half the number of bichrome V's .

Finally, we compute an upper bound for the number of bichrome V's . We note that from the 5 edges meeting at a node $0*5$ or $1*4$ or $2*3$, i.e. at most 6 , bichrome V's meeting at that node can be constructed. Hence the total number of bichrome V's is at most $6 * 6$, the total number of bichrome triangles is at most $6 * 6/2$, and hence the total number of monochrome triangles is at least 2.

End Exposition1 .

$$*$$

The "invention" enabling us to construct the above argument is, of course, the notion of a bichrome V . The concept, however, does not come out of the blue: it is the result of our having realized the option of counting monochrome triangles by counting bichrome ones and of the decision to keep things symmetric and simple. None of these circumstances should be surprising. The bichrome V effectively hides the individual colours —and rightly so, because their only rôle is to express equality and difference of colour— in very much the same way as in Chapter 1, the x and the y hide individual bits by standing for a pair of different and equal neighbour bits respectively. In this respect we note that two edges of different colour form the simplest ensemble that is invariant under colour inversion.

3 Proving the existence of the Euler line

This chapter is concerned with some consequences of introducing nomenclature: repetitiousness, caused by the destruction of symmetry that is inherent to giving different things different names, and lack of disentanglement, caused by the availability of avoidable nomenclature. A second point the chapter wants to illustrate —and remedy— is how the use of pictures has the danger of strongly inviting (i) the introduction of too much nomenclature, and (ii) implicitness about the justification of the steps of the argument.

In the following we present two proofs of the existence of the Euler line, a standard argument in Exposition0 and an alternative argument in Exposition1.

Theorem . The orthocentre, the centroid, and the circumcentre of a triangle all lie on a single line: the (not necessarily unique) Euler line.

Exposition0 .

Proof . Let H be the orthocentre, G the centre of gravity, and M the circumcentre of triangle ABC . Multiply the whole figure with respect to G with a factor $-1/2$, so that C is mapped onto the midpoint of AB and cyclically A onto the midpoint of BC and B onto the midpoint of CA . Of course the images C' , A' , and B' respectively are such that $A'B'//AB$, etc., so that M is the orthocentre of triangle $A'B'C'$, or $M = H'$.

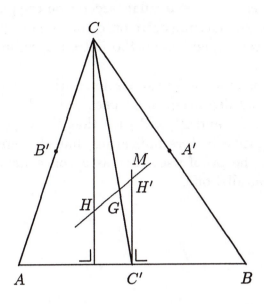

End Exposition0 .

Exposition1 .   In a triangle

(0)   the perpendicular bisector and the altitude of a side are parallel; the first goes through the midpoint of the side, the second through the opposite vertex;

(1)   the median of a side connects the midpoint of the side with the opposite vertex;

(2)   the three perpendicular bisectors concur in the circumcentre, the altitudes in the orthocentre, and the medians in the centroid,

(3)   the latter dividing each median in the same ratio (viz. $1 : 2$).

Proof of Theorem .   There exists a multiplication (viz. with a factor -2) with respect to the centroid that

> maps each midpoint of a side onto the opposite vertex,
> by (1) and (3) ,   and
> transforms each line into a parallel one,

hence, by (0), maps each perpendicular bisector on the parallel altitude, hence, by (2), maps the circumcentre onto the orthocentre; hence the centroid, the circumcentre, and the orthocentre are collinear.

This proof uses that a multiplication with a factor m with respect to a point X by definition maps each point P onto a point Q that is collinear with X and P and such that the proportion of XQ and XP is m (negative factors indicating that X separates a point and its image), and the proof uses that as a consequence each line is transformed into a parallel one.

End Proof.

End Exposition1.

*   *
*

Both proofs seem of roughly the same length, but are only superficially so, since Exposition0 leaves quite a few things implicit:

(i)   It does not mention the convention of using primes to distinguish the image from the original; it only uses it.

(ii)   It does not define the notions circumcentre, centroid, and orthocentre; it only uses them.

(iii)   Neither does it define the notions perpendicular bisector, median, and altitude; what is more, they are not even mentioned.

(iv)   It does not mention the theorem that the centroid divides each median in a ratio $1 : 2$; it only uses it.

(v)   It takes for granted that multiplication with respect to a point is such that when A, B, and C are mapped onto A', B', and C' respectively, the orthocentre of triangle ABC is mapped onto the orthocentre of triangle $A'B'C'$.

Circumstances (iv) and (v) are an immediate consequence of (iii): without the notions perpendicular bisector, median, and altitude and their properties, the terminology is lacking to formulate and justify the theorems in (iv) and (v) in any detail.

Although on the one hand being quite implicit, Exposition0 on the other hand is quite overspecific. It introduces the names H, G, and M, which not counting the picture are used about once. It names all the vertices of the triangle and as a result shows that naming can destroy symmetry: the symmetry has to be saved by enumeration or its "substitute", the elliptical "etc.".

In Exposition1 we decided to try to leave the vertices anonymous. Then, not being able to distinguish easily between vertices (or sides for that matter) we were more or less straightforwardly led to concentrate on *one* side and its altitude, median, and perpendicular bisector. As a result, we arrived at an argument in which the fact that three sides and vertices are involved only enters the discussion by means of the perfectly symmetric notions circumcentre, centroid, and orthocentre. This is in contrast with Exposition0, where in the expression $A'B'//AB$ at least two sides are involved.

Having formulated our argument in a satisfactory degree of detail without the introduction of names, we are able to follow Lagrange by dispensing with the picture: it has proved to be superfluous. The picture, however, is not only superfluous, but misleading as well: does the theorem hold for an obtuse triangle? In presentations like these the best a picture can do is to give an example, an instance. As such it is eo ipso overspecific. It invites, almost forces, the reader to deal with the general case in terms of a specific instance, the particulars of which are to a large extent left implicit.

4 In adherence to symmetry

This chapter is another illustration of the complications engendered by the introduction of nomenclature, here emerging in the form of overspecificity and loss of symmetry. It also discusses the choice between recursion and complete unfolding.

We consider couplings, i.e. one-to-one correspondences, between two equally sized finite bags of natural numbers. Hence, a coupling can be considered a bag of —ordered— pairs of numbers, the subbags of which are as usual called its subcouplings. The value of a coupling is defined recursively by

–the value of an empty coupling is 0 ;
–the value of a one-element coupling is the product of the members in the single pair;
–the value of a non-empty coupling is the value of one element + the value of the remaining subcoupling.

Note . By the associativity and symmetry of addition, the above is a valid definition.
End Note .

The problem is to construct a coupling with maximum value. Such a maximum value exists, since the finite bags have a finite number of couplings.

A construction follows from the two lemmata

Lemma0 .   Each subcoupling of a maximum coupling is itself a maximum coupling.

Lemma1 .   In a maximum non-empty coupling, the maximum values of the two bags form a pair.

The construction then consists in choosing the maximum values to form a pair and constructing a maximum coupling for the remainders of the bags in the same way.   The construction terminates since the bags are finite and decrease in size at each step.

Proof of Lemma0 . By the symmetry and associativity of addition we have —with $\cup$ for bag union—

> the value of coupling $B \cup C$ = the value of B + the value of C ;

Lemma0 now follows from the monotonicity of addition.

Proof of Lemma1 . We consider a maximum coupling, in which the maximum values U and V of the bags being coupled are paired with v and u respectively, and we prove $v = V \lor u = U$.

- If (U, v) and (u, V) are the same element of the coupling, $U = u \land v = V$.
- If (U, v) and (u, V) together form a two-element subcoupling we have

>   **true**
> $=$ {by Lemma0 and maximality of the coupling}
>   value of $\{(U, v), (u, V)\}$ $\geq$ value of $\{(U, V), (u, v)\}$
> $=$ {definition of "value"}
> $U * v + u * V \geq U * V + u * v$
> $=$ { }
> $(U - u) * (v - V) \geq 0$
> $=$ { $(U - u) * (v - V) \leq 0$, since $U \geq u \land V \geq v$ }
> $(U - u) * (v - V) = 0$

$$= \quad \{\ \} $$
$$U = u \ \lor \ v = V \quad .$$

End Proofs .

$$* \qquad\qquad *$$
$$*$$

The above theorem is far from new, but in the literature one usually finds a rather different formulation, which is essentially as follows.

> Given an ascending sequence a_i , $0 \le i < N$, of natural numbers and a sequence b_i of natural numbers, we are requested to maximize $(\underline{S} i : 0 \le i < N : a_i * b_{\pi(i)})$, where π ranges over the permutations of $i : 0 \le i < N$.

There is quite some overspecificity in this formulation. Firstly, instead of introducing bags it introduces sequences, as a result of which permutations inevitably enter the problem statement. (In "Inequalities" by Hardy, Littlewood, and Pólya, for instance, the problem is dealt with under the heading "Rearrangements".) Any ordering of elements, however, is foreign to the problem of finding some one-to-one correspondence and should therefore only be introduced with care and good reason. In the above, sequences are probably used as a generalization of sets allowing multiple occurrences of values; the bag, however, by not involving order, is a more appropriate generalization.

Next, the above formulation introduces names for all the elements of the sequences, consequently also for the length of the sequences and for the permutation. Apparently none of these names is necessary.

Finally, the ascendingness of sequence a_i is irrelevant for the value of the sum to be maximized, but its introduction immediately destroys the symmetry between the two sequences. This we consider a disadvantage.

In summary, the above observations led us to head for an argument phrased in terms of bags, being completely symmetric in the bags, and using as few names as necessary. A consequence of not naming the elements was the obligation to define the value to be maximized, i.e. the value of a coupling. This, however, straightforwardly led us to Lemma0 and hence to the recursive description of the construction comprised by Lemma0 and Lemma1.

We note that at the few places in the argument where we introduced names —most notably in the proof of Lemma1— we have chosen them so as to reflect the symmetry between the bags.

Finally we note that the ascendingness of sequence a_i and permuted sequence b_i in the traditional formulation can be viewed as the completely unfolded version of our recursive scheme. This choice between recursion and complete unfolding is worth noting; recursive formulations tend to be more compact than their unfoldings and, hence, may be more convenient to manipulate.

(Another illustration is the proof of a probably well-known theorem relating the number of times a prime p divides into $n!$ to the sum of the digits of n's p-ary representation, viz. with $f.n$ being the former and $s.n$ the latter, the theorem says $f.n = (n - s.n)/(p - 1)$.

The recursive definition of $n!$ immediately suggests a proof by induction over n. (See EWD538 in "Selected Writings on Computing: A Personal Perspective" by Edsger W. Dijkstra.) The recursive definition may even assist in deriving the theorem itself. The unfolded definition of $n!$ $-n! = (\underline{P} i : 1 \leq i \leq n : i)-$ provides much more possibilities for massaging and, hence, is more difficult from a heuristic point of view. In books on number theory its use, for instance, leads to $f.(n) = n \underline{\text{div}} p + n \underline{\text{div}} p^2 + \cdots + n \underline{\text{div}} p^k + \cdots$, a relation that does not lend itself as easily for manipulation or for proof by induction over n as the earlier one.)

5 On a proof by Arbib, Kfoury, and Moll

> This chapter discusses an extreme example of the harm done
> by the introduction of nomenclature that forces the making
> of avoidable distinctions, in particular the introduction of sub-
> scripted variables. In addition it illustrates some consequences
> of neglecting equivalence as a connective in its own right.

The proof we discuss below takes to the extreme some unfortunate
though common mathematical habits. It is taken from "A Basis for Theoretical Computer Science" by Michael Arbib, A.J. Kfoury, and Robert
N. Moll. That the authors consider their proof exemplary illustrates
a widespread neglect of the circumstance that for computing scientists,
perhaps even more than for mathematicians in general, effective mathematical arguments are not a luxury but often a sheer necessity for keeping
complexity at bay.

One of the habits alluded to above is proving the equivalence of
two statements by proving mutual implication. We have noticed that
for many mathematicians following this pattern of proof has become
second nature, even so much so that some advocate it as the only pattern
possible. The authors mentioned above, for instance, are convinced
that

> "to prove $A \Leftrightarrow B$ [...] we must actually complete two
> proofs [...] : We must prove both $A \Rightarrow B$ and its
> converse $B \Rightarrow A$." .

The exemplary proof with which they want to demonstrate their point
is a proof of

(0)   "Theorem.   A natural number is a multiple of 3 iff the
        sum of the digits in its decimal representation is a mul-
        tiple of 3 ."   .

Their proof by mutual implication, however, is self-inflicted: it is solely
due to the shape of the auxiliary theorem from which (0) is derived.
That theorem reads, with $r(n)$ being defined as the sum of the digits of
natural number n's decimal representation,

(1)   "If $n = 3m$, then $r(n)$ is a multiple of 3 .
        If $n = 3m + 1$, then $r(n)$ is of the form $3k + 1$.
        If $n = 3m + 2$, then $r(n)$ is of the form $3k + 2$."   .

We can only guess why the authors have not chosen a concise formulation
without cases and with a minimum amount of nomenclature, such as

$$r(n) \bmod 3 \qquad = n \bmod 3 \qquad , \text{ with } (0) \text{ as the special case}$$
$$r(n) \bmod 3 = 0 \quad \equiv n \bmod 3 = 0 .$$

Perhaps they were reluctant to use the extra concept $\bmod$. Another
possible explanation is that they have chosen formulation (1) because
it reflects most directly the proof they had in mind: (without actually
giving the proofs) the authors say that (1), in its turn, is to be derived
from

(2)   "$r(n + 3)$ differs from $r(n)$ by a multiple of 3 ."   ,

"by induction on m for each of the three cases". Apparently they
force themselves to make do with the restricted form of mathematical
induction that allows steps from m to $m + 1$ only. Indeed, in their
treatment of induction they write:

    "First, we check that the property holds for 0 . [...]
    Then we prove that whenever any n in N satisfies the
    property, it must follow that $n + 1$ satisfies the property.
    [...] Proofs by induction may start at 1 as well as 0 ;

and, indeed, proofs of this kind may start at any positive integer." .

Such a restricted form of mathematical induction is, of course, a self-inflicted pain. It is inadequate for mathematicians in general, and even more so for computing scientists who need, in termination arguments for programs, various well-founded sets besides the natural numbers.

Finally, the proof of (2) presented by Arbib, Kfoury, and Moll is an extreme example of the harm done by making too many distinctions, especially if it is combined with a lot of overspecific nomenclature. In that proof, when describing how the decimal representation of $n + 3$, denoted by $\langle n + 3 \rangle$, can be derived from the decimal representation of n, denoted by $\langle n \rangle$, the authors distinguish between 0 carries and at least 1 carry, subdividing the latter into (i) exactly 1 carry, (iii) as many carries as $\langle n \rangle$ has digits, and (ii) a number of carries in between, and once more subdividing subcase (i) according to whether $\langle n \rangle$ has 1 or more digits, i.e. according to whether the case overlaps with (iii). (The proof is shown further on in this note.)

So not only do the authors distinguish between 0 and the other natural numbers, they also treat 1 differently from the larger natural numbers. In making these distinctions, however, they are in no way exceptional. Hardy and Wright's very first theorem in "An introduction to the theory of numbers", for instance, reads: "Every positive integer, except 1, is a product of primes", thus neglecting 1 as a product of 0 primes; and Harold M. Stark, in "An Introduction to Number Theory", writes: "If n is an integer greater than 1, then either n is prime or n is a finite product of primes", thus not only neglecting 1 as a product of 0 primes, but also neglecting that a product of 1 prime is a finite product of primes. Because both books are still widely used, the distinctions continue to be made.

We claim that such distinctions as the present authors make are the result of introducing nomenclature. By introducing

$$d_m d_{m-1} \ldots d_1 d \quad (m \geq 0)$$

for $\langle n \rangle$, they force themselves to express $\langle n+3 \rangle$ in the same terminology. Consequently, they have to worry about which digit values change, which index values these digits have, and even about whether the length of the digit sequence changes.   This leads them to their complicated proof of theorem (2) , which we add for the sake of completeness and reference.

"Proof.   We prove the result by exhausting two cases:

(I)   The last digit d of $\langle n \rangle$ is $0, 1, 2, 3, 4, 5$ or 6. In this case we form $\langle n + 3 \rangle$ by changing d to $d + 3$. Thus, $r(n + 3) = r(n) + 3$, satisfying the claim of the lemma.

(II)   The last digit d of $\langle n \rangle$ is $7, 8$, or 9. In that case we form $\langle n+3 \rangle$ from the string $\langle n \rangle = d_m d_{m-1} \ldots d_1 d(m \geq 0)$ of digits by the following rule, which exhausts three possible subcases:

(1)   If $d_1 \neq 9$, set $\langle n+3 \rangle = d_m d_{m-1} \ldots (d_1 + 1)(d-7)$. (If $m = 0$, this rule changes $\langle n \rangle$ to $1(d-7)$.)   Then $r(n + 3) = 1 + r(n) - 7 = r(n) - 6$, satisfying the claim of the lemma.

(2)   If $\langle n \rangle = d_m d_{m-1} \ldots d_{k+2} d_{k+1} 9 \ldots 9d$ with $d_{k+1} \neq 9$ (where $1 \leq k \leq m$), set
$\langle n + 3 \rangle = d_m d_{m-1} \ldots d_{k+2}(d_{k+1} + 1)0 \ldots 0(d-7)$.   Then $r(n + 3) = r(n) - 9k - 6$, satisfying the claim of the lemma.

(3)   If $\langle n \rangle = 9 \ldots 9d$, set $\langle n+3 \rangle = 10 \ldots 0(d-7)$.   Then $r(n+3) = r(n) - 9m - 6$, satisfying the claim of the lemma. Having verified the lemma for all subcases, we have proved it to be true.   □"

*   *

*

There are, of course, many effective ways to prove (0) , such as the following.   It proves $r(n) = n + $ multiple of 9 .   We adopt the decimal positional system, dropping, however, the constraint that each digit be less than 10. We start with the decimal number having n as its only significant "digit" —and, hence, having digit sum n— , and by repeated carries transform it into n's standard decimal representation —which has digit sum $r(n)$— . A carry consists in subtracting 10 from a digit ≥ 10 and adding, or creating, 1 at the position to the immediate left. Each carry, hence, changes the digit sum by -9. The process terminates because the digit sum is at least 0 and decreases at each carry. Hence $r(n) = n + $ multiple of 9.

6 Not about open and closed sets

This chapter is primarily included as an example of orderly and explicit proof development guided by the shape of the formulae rather than by their interpretation. In passing it illustrates the usefulness of the equivalence in massaging proof obligations. In revealing the structure of our argument clearly and in justifying in a concise way why each step is taken, the use of formalism is essential.

In the following we construct a proof for a theorem that occurs in most elementary books on analysis. Tom M. Apostol, for instance, in "Mathematical Analysis A modern Approach to Advanced Calculus", formulates it as follows, for a subset S of some universe E_1 ,

"Theorem. If S is open, then the complement $E_1 - S$ is closed. Conversely, if S is closed, then $E_1 - S$ is open."

(The reader that considers the above to be a definition rather than a theorem is invited to view our proof as a proof of equivalence of two definitions. For what follows the distinction is irrelevant.)

First of all, because we have learned to appreciate equivalence and symmetry so much, we prefer to formulate the theorem differently, viz.

Theorem . Subsets S and T of some universe that are each other's complement satisfy

S is open $\equiv T$ is closed   .

End Theorem .

Next the only thing we can do is to consider the definitions of "complement", "open", and "closed". Firstly we have

(0)   subsets S and T of some universe are each other's complement means $(\underline{A}\,x :: x \in S \not\equiv x \in T)$.

Following Apostol, we next define

(1) S is open $\equiv$ $(\underline{A}\,x : x \in S : iS.x)$
(2) T is closed $\equiv$ $(\underline{A}\,x : aT.x : x \in T)$,

where $iS.x$ means "x is an interior point of S" and $aT.x$ means "x is an accumulation point of T", notions to be detailed later.

The right-hand sides of (1) and (2) show a similarity in structure that we can even increase by rewriting (2)'s right-hand side: replacing $x \in T$ by $\neg(x \in S)$, on account of (0), and by predicate calculus, we get

$$RHS(2) \equiv (\underline{A}\,x : x \in S : \neg aT.x)   ;$$
$$RHS(1) \equiv (\underline{A}\,x : x \in S : iS.x)   .$$

So we can prove the theorem, the equivalence of $RHS(2)$ and $RHS(1)$, by proving

(3) $\neg aT.x \equiv iS.x$ for $x \in S$.

Remark . Proving the equivalence of two universal quantifications by proving the equivalence of their terms means strengthening the demonstrandum quite a bit; it is, however, the simplest step suggested by the formulae and, therefore, always worth being investigated.
End Remark .

Now the only thing we can do in order to prove (3) is to consider the definitions of $iS.x$ and $aT.x$, i.e. of the notions interior point and accumulation point. With dummy V ranging over neighbourhoods of x we have, again following Apostol,

$$aT.x \;\equiv\; (\underline{A}\,V :: (V-\{x\}) \cap T \neq \phi) \;,$$

i.e. by De Morgan's Law,

$$\neg aT.x \;\equiv\; (\underline{E}\,V :: (V-\{x\}) \cap T = \phi) \;\; ; \text{ and}$$
$$iS.x \;\equiv\; (\underline{E}\,V :: V \subseteq S) \;.$$

Again, (3) can be proved by proving the equivalence of the terms of the above two —existential— quantifications, i.e. by proving

(4) $(V-\{x\}) \cap T = \phi \;\equiv\; V \subseteq S$
        for $x \in S$ and V a neighbourhood of x .

This we do as follows:

$$(V-\{x\}) \cap T = \phi$$
$$= \qquad \{T \text{ and } S \text{ are complements}\}$$
$$V-\{x\} \subseteq S \qquad\qquad\qquad\qquad\qquad *$$
$$= \qquad \{x \in S \,,\ \text{see } (3)\}$$
$$(V-\{x\}) \cup \{x\} \subseteq S$$
$$= \qquad \{\text{set calculus}\}$$
$$V \cup \{x\} \subseteq S$$
$$\Rightarrow \qquad \{\text{set calculus}\}$$
$$V \subseteq S$$
$$\Rightarrow \qquad \{\text{set calculus}\}$$
$$V-\{x\} \subseteq S \quad . \qquad\qquad\qquad\qquad *$$

By the recurrence of expression $*$, all expressions in the calculation are equivalent, in particular the first and the one but last, hence (4) has been established.

Aside . We have proved (4) by means of set calculus because the demonstrandum presented itself in terms of sets. We wish to note, however, that a proof in terms of the corresponding characteristic predicates, by means of predicate calculus, can be a nicer alternative. When phrased as

$$[\neg(V \wedge \neg X \wedge T)] \equiv [V \Rightarrow S] \quad , \text{given } [X \Rightarrow S] \text{ and } [S \not\equiv T] ,$$

(4) can, for instance, much more easily be proved by means of equivalence transformations only. (As usual in this study, square brackets denote universal quantification over the universe.) Such a proof might run as follows

$$\neg(V \wedge \neg X \wedge T)$$
$$= \qquad \{\text{De Morgan}\}$$
$$\neg V \vee X \vee \neg T$$
$$= \qquad \{[S \equiv \neg T], \text{ since } [S \not\equiv T]\}$$
$$\neg V \vee X \vee S$$
$$= \qquad \{[X \vee S \equiv S], \text{ since } [X \Rightarrow S]\}$$
$$\neg V \vee S$$
$$= \qquad \{\text{implication}\}$$
$$V \Rightarrow S \quad .$$

Hence we have proved the stronger:

$$[\neg(V \wedge \neg X \wedge T) \equiv (V \Rightarrow S)] \quad .$$

End Aside .

This completes the proof.

$$* \qquad \qquad *$$
$$*$$

We have presented the above proof because we wanted to show that its design is a purely syntactic affair. We started by writing down formal definitions of open and closed, and from a syntactic analysis of these definitions we immediately derived the next step of the proof. Then we did exactly the same once more, reducing the demonstrandum to a relation between sets, and that was all. No "invention" involved at all.

Note that with the exception of (0) each definition is used only once in the proof. This we consider to be indicative of an effective separation of concerns. The only property of neighbourhoods used is their being sets. For the proof it is irrelevant whether a point belongs to its neighbourhoods. That (0) is used only twice means that for the larger part of the argument this relation between S and T is irrelevant. Hence introducing a separate name T rather than using $E_1 - S$ had a few advantages. We could avoid frequent repetition of an expression whose internal structure is largely irrelevant but that lengthens and complicates the formulae and their parsing. Furthermore we did not have to introduce a name for the universe.

$$* \qquad\qquad *$$
$$*$$

By way of contrast we add Apostol's proof of the theorem. To simplify the comparison we repeat our proof, leaving out all heuristic remarks.

S is open
$=$ $\{$definition of open$\}$

$(\underline{A} x : x \in S : iS.x)$
$=$ $\{ iS.x \equiv \neg aT.x$ for $x \in S$, see below$\}$

$(\underline{A} x : x \in S : \neg aT.x)$
$=$ $\{$predicate calculus and $(0)\}$

$(\underline{A} x : aT.x : x \in T)$
$=$ $\{$definition of closed$\}$

T is closed   .

$$\neg aT.x$$
$$= \quad \{\text{definition of } aT.x \}$$
$$\neg(\underline{A} V :: (V - \{x\}) \cap T \neq \phi)$$
$$= \quad \{\text{De Morgan}\}$$
$$(\underline{E} V :: (V - \{x\}) \cap T = \phi)$$
$$= \quad \{\text{see } (4)\}$$
$$(\underline{E} V :: V \subseteq S)$$
$$= \quad \{\text{definition of } iS.x \}$$
$$iS.x \quad .$$

The proof of (4) is as before.

Apostol's proof.

"3-15 THEOREM. If S is open, then the complement $E_1 - S$ is closed. Conversely, if S is closed, then $E_1 - S$ is open.

Proof. Assume that S is open and let x be an accumulation point of $E_1 - S$. If $x \notin E_1 - S$, then $x \in S$ and hence some neighborhood $N(x) \subset S$. But, being a subset of S, this neighborhood can contain no points of $E_1 - S$, contradicting the fact that x is an accumulation point of $E_1 - S$. Therefore, $x \in E_1 - S$ and $E_1 - S$ is closed.

Next, assume that S is closed and let $x \in E_1 - S$. Then $x \notin S$ and hence x cannot be an accumulation point of S, since S is closed. Therefore some neighborhood $N(x)$ has no points of S and must consist only of points of $E_1 - S$. That is, $N(x) \subset E_1 - S$ and $E_1 - S$ must therefore be open."

End Apostol's proof .

The two proofs have roughly the same number of symbols (about

500), but the former is much more explicit, about definitions, about the justification of steps, and about the structure of the argument.   We note that Apostol has obscured the similarities between the two parts of his proof in three ways: by not formulating the second conjunct of his theorem as "If $E_1 - S$ is closed, then S is open"; by avoiding the concept "interior point" while using the concept "accumulation point"; and by using a (superfluous) proof by contradiction in only one of the two parts.

7 A monotonicity argument

The belief that equivalence is always most appropriately proved by showing mutual implication has undoubtedly been strengthened by the way in which proofs in Euclidean geometry are conducted. The purpose of this chapter is to show that some of that "geometrical evidence" is not compelling at all.

We consider a function f of two arguments satisfying

(0) $[x < y \Rightarrow f.x.y < f.y.x]$,

and show that it satisfies

(1) $[x = y \equiv f.x.y = f.y.x]$.

Note . Square brackets denote universal quantification over x and y . Variables x and y are taken from a universe on which a total order relation " $<$ " is defined; similarly for the range of function f .
End Note .

Proof .

$\qquad$ (0)

$=\qquad$ {by definition}

$\qquad [x < y \Rightarrow f.x.y < f.y.x]$

$=\qquad$ {doubling, once renaming the dummies}

$\qquad [x < y \Rightarrow f.x.y < f.y.x] \wedge [y < x \Rightarrow f.y.x < f.x.y]$

41

$\Rightarrow \qquad \{$term-wise disjunction$, p < q \lor q < p \equiv p \neq q\}$

$\qquad [x \neq y \Rightarrow f.x.y \neq f.y.x]$

$= \qquad \{$contrapositive$\}$

$\qquad [x = y \Leftarrow f.x.y = f.y.x]$

$= \qquad \{[x = y \Rightarrow f.x.y = f.y.x]\}$

$\qquad [x = y \equiv f.x.y = f.y.x]$

$= \qquad \{$by definition$\}$

$\qquad (1) \qquad .$

End Proof .

Remark . In addition to (1), other equivalences can be derived from (0). Their derivations are, however, deferred to an appendix, because they are irrelevant for this chapter's main topic.

End Remark .

*

In Euclidean geometry the two theorems

- "An isosceles triangle has two equal angles",   and
- "A triangle with two equal angles is isosceles"

are usually proved separately, with congruences. Usually prior to that it is established that opposite to the larger angle lies the larger side, in the jargon: $\alpha < \beta \Rightarrow a < b$. For fixed "third side" c, this is a statement of form (0); so congruences are not needed for the proofs of the two theorems, since the latter are subsumed in the conclusion corresponding to (1): $\alpha = \beta \equiv a = b$.

The same holds for the two theorems

- "An isosceles triangle has two angle bisectors of equal length" and
- "A triangle with two angle bisectors of equal length is isosceles" .

Thanks to all the above, both theorems are subsumed in

(2)   "In a triangle, the larger angle has the shorter angle bisector"   ,

which, again for fixed base of the triangle, is also of form (0) .

Theorem (2) can be proved in various ways. For instance, in a triangle with sides of length a, b, and c, the square of the length of the angle bisector of α, the angle opposite to side a, equals

(3) $b * c * (1 - (a/(b + c))^2)$,

an expression that is increasing in b and decreasing in a and, therefore, satisfies (0): an f that is increasing in its first argument and decreasing in its second, satisfies

$$[x < y \Rightarrow f.x.y < f.y.y \land f.y.y < f.y.x]   .$$

So much for theorem (2). (As for (3), it follows by applying the cosine rule in the two subtriangles into which the original triangle is divided by the angle bisector, and eliminating the equal cosines.) For the sake of completeness we note that the whole argument pertains to interior angle bisectors only.

*   *

*

Appendix . In addition to implying equivalence (1) , viz. $[x = y \equiv f.x.y = f.y.x]$, (0) also implies

(4) $[x \leq y \equiv f.x.y \leq f.y.x]$ and
(5) $[x < y \equiv f.x.y < f.y.x]$.

Firstly, (5) $\equiv$ (4) :

$$(5)$$
$$=   \{ \}$$
$$[x < y \equiv f.x.y < f.y.x]$$

$=$ {negation of both sides, $\neg(p < q) \equiv q \leq p$}

$[y \leq x \equiv f.y.x \leq f.x.y]$

$=$ {renaming the dummies}

(4) .

Relation (4) is proved by proving mutual implication:

$[x \leq y \Rightarrow f.x.y \leq f.y.x]$

$\Leftarrow$ {term-wise disjunction, using $[x = y \Rightarrow f.x.y = f.y.x]$}

$[x < y \Rightarrow f.x.y < f.y.x]$

$=$ { }

(0)

and

$[x \leq y \Leftarrow f.x.y \leq f.y.x]$

$=$ {contrapositive, $\neg(p \leq q) \equiv q < p$}

$[y < x \Rightarrow f.y.x < f.x.y]$

$=$ {renaming the dummies}

(0) .

Finally we note that (1), (4), and (5) also hold for the special case that f depends on only one of its arguments, i.e. if f is an increasing function or a decreasing function.

End Appendix .

This chapter is a revised version of technical report AvG36/EWD878, with Edsger W. Dijkstra.

8 On the inverse of a function

This very small essay tracks down the origin of an asymmetry in the usual treatments of the notion of the inverse of a
function, and does away with that asymmetry. It is another
exercise in maintaining equivalence.

For a binary relation R on a Cartesian product $A \times B$ we define

$$
\begin{array}{llll}
R \text{ is total} & \equiv & (0)\text{:} & (\underline{A}\,x : x \in A : (\underline{N}\,y : y \in B : xRy) \geq 1) \\
R \text{ is surjective} & \equiv & (1)\text{:} & (\underline{A}\,x : x \in B : (\underline{N}\,y : y \in A : yRx) \geq 1) \\
R \text{ is functional} & \equiv & (2)\text{:} & (\underline{A}\,x : x \in A : (\underline{N}\,y : y \in B : xRy) \leq 1) \\
R \text{ is injective} & \equiv & (3)\text{:} & (\underline{A}\,x : x \in B : (\underline{N}\,y : y \in A : yRx) \leq 1) \quad .
\end{array}
$$

The symmetry exhibited by (0)–(3) is such that $(1) = (0)(A, B, xRy := B, A, yRx)$ and, similarly, $(3) = (2)(A, B, xRy := B, A, yRx)$. It is re-
flected in the notion "converse":

> relation R on $A \times B$ and relation S on $B \times A$
> are each other's converse
>
> $\equiv$
>
> $(\underline{A}\,x, y : x \in B \wedge y \in A : xSy \equiv yRx)$; hence

$(1) = (0)(A, B, R := B, A, S)$ and $(3) = (2)(A, B, R := B, A, S)$, i.e.

$$
\begin{array}{lll}
(4) & R \text{ is surjective} & \equiv S \text{ is total} \\
(5) & R \text{ is injective} & \equiv S \text{ is functional} \quad .
\end{array}
$$

For relations R and S that are each other's converse we define

45

S is the inverse of R $\equiv$ R is functional $\wedge$ S is functional ,

the symmetry of which immediately yields

S is the inverse of R $\equiv$ R is the inverse of S .

Finally, by (5) we have

R's converse is R's inverse $\equiv$ R is functional and injective .

(One may prefer to confine the notion of an inverse to total relations; this is incorporated by conjoining the right-hand side of the definition with "R is total $\wedge$ S is total", which by (4) is equivalent to "R is total and surjective".)

*   *
*

The above grew out of the following observation. In many textbooks the introduction of the inverse of a function terminates with a proof of the theorem "If f is the inverse of g then g is the inverse of f." . Apparently the symmetry of the relation "being each other's inverse" requires proof. Since in these treatments the notion of an inverse is based on the symmetric notion of the converse, just like it is in the above treatment, that means that somewhere along the way from converses to inverses the symmetry is destroyed.

In the proofs (see, for instance, "Mathematical Analysis, A Modern Approach to Advanced Calculus" by Tom M. Apostol) the symmetry is primarily broken because functionality and injectivity are not treated symmetrically: the latter notion is confined to functions. However, the notation $R(x)$, introduced for the unique solution y of xRy in the case of a functional relation, is less suitable for rendering relations like the above.

As a result, we prefer to attach all four notions to relations rather than functions. If one then defines the notions formally in as concise and homogeneous a way as possible, the rest of the treatment becomes hardly more than a simple syntactic activity. We note that the use of equivalences wherever possible has smoothened this activity.

9 A calculational proof of Helly's theorem on convex figures

The proof in this chapter is included firstly to show the calculational style in action, this time in a geometrical problem, and, secondly, to illustrate the carefully phased exploitation of data that is enabled by the introduction of nomenclature.

Preliminary remark . In this chapter, we aim at illustrating how the shape rather than the interpretation of the formulae that emerge leads the way in the design of a proof. For that purpose, the proof proper is interspersed with considerations of heuristic flavour. Wherever the distinction could be made, we have isolated these heuristic considerations in Notes.
End Preliminary remark .

Helly's theorem is concerned with finite sets of figures, "tuples" for short, for which the following notions are relevant:

a figure is a set of points, and for tuple t
$B.t \equiv$ the figures of t have a point in common
$C.t \equiv$ each 3 figures of t have a point in common
$\#t =$ the number of figures in t .

Helly's theorem . A tuple t of plane convex figures such that $3 \le \#t$, satisfies $C.t \Rightarrow B.t$.
End Helly's theorem .

The proof given below consists of 3 sections and an appendix.

• The initial section eliminates concept C from the demonstrandum; it does so by the use of mathematical induction. (In passing, the constant 3 all but disappears from the demonstrandum, thus paving the way for a generalization of the theorem.)

• The next section takes into account that B has the form of an existential quantification over a dummy of type point. The proof obligation in terms of B is rephrased in terms of a predicate D defined on points.

• The third section meets the final proof obligation in terms of D by taking the convexity of the figures and the dimension of the space into account.

• The appendix contains deferred proofs.

$$* \qquad *$$
$$*$$

The proof. In the following, identifiers $r, s, t,$ and p denote tuples, i.e. sets of figures. For later use we first note that B and C are monotonic with respect to tuple containment " $\supseteq$ " , i.e.

$$r \supseteq s \quad \Rightarrow \quad (B.r \Rightarrow B.s) \wedge (C.r \Rightarrow C.s) \quad .$$

Note0 . For the removal of C from the demonstrandum, we investigate what proof obligations emerge if we attempt to prove Helly's theorem by induction on $\#t$.
End Note0 .

Induction on $\#t$:

• Base: $3 = \#t$ implies $C.t \equiv B.t$, since t is the one and only subtriple of t .

• Step: we wish to derive, for $3 < \#t$, $C.t \Rightarrow B.t$ from induction hypothesis $(\underline{A} p : 3 \leq \#p < \#t : C.p \Rightarrow B.p)$. We calculate

$C.t$

$\Rightarrow$ {monotonicity of C w.r.t. $\supseteq$ }

$(\underline{A}\,p : t \supseteq p \,\wedge\, 3 \leq \#p < \#t : C.p)$

$\Rightarrow$ {induction hypothesis, monotonicity of $\underline{A}$ }

$(\underline{A}\,p : t \supseteq p \,\wedge\, 3 \leq \#p < \#t : B.p)$

$\Rightarrow$ { ? , heading for $B.t$ }

$B.t$,

a calculation whose last as yet unjustified step suggests us to choose that step as our remaining proof obligation, viz.

$$B.t \,\Leftarrow\, (\underline{A}\,p : t \supseteq p \,\wedge\, 3 \leq \#p < \#t : B.p)$$

or, by B's monotonicity w.r.t. $\supseteq$ equivalently,

(i) $\qquad B.t \,\Leftarrow\, (\underline{A}\,p : t \supseteq p \,\wedge\, \#p = \#t - 1 : B.p)$ for $3 < \#t$.

Note1 . We omit "$3 \leq \#p$" from (i) because it follows from $\#p = \#t - 1 \,\wedge\, 3 < \#t$. We note that with C's disappearance from the demonstrandum, nearly all traces of the specific "**3**" have disappeared. We also note that so far the only property about "**3**" used is the equality of $\#t$ for the base of the induction and the "**3**" in $C.t$; replacing all symbols 3 by k , hence, maintains the validity of the argument.
End Note1 .

*   *

Having removed C , we now concentrate on B . In order to prove (i) , we shall have to take more details of B into account. More formally, for r a tuple of figures,

$$B.r \,\equiv\, (\underline{E}\,Q :: (\underline{A}\,f : f \,\epsilon\, r : Q \text{ in } f)) \quad ,$$

where Q is a dummy of type point and "Q in f" stands for "point Q is a point of figure f" . From here onwards, dummies r, s , and p will be confined to subtuples of t .

Drawing from our experience in programming, we isolate the inner quantification of $B.r$, by naming it, and investigate its properties:

$$[D.r \equiv (\mathbf{A} f : f \epsilon t \backslash r : Q \text{ in } f)] \quad .$$

Notational remark . In the above, square brackets denote universal quantification over Q —note that $D.r$ implicitly depends on Q— and $t \backslash r$ denotes r's complement in t, i.e. $f \epsilon t \backslash r \equiv f \epsilon t \wedge \neg (f \epsilon r)$; so $D.r$ also implicitly depends on t . We have only made the dependence on r explicit, because it is the only parameter in $D.r$ for which we shall need different instantiations.
End Notational remark .

The purpose of D's introduction is to remove B . We have, for any tuple r , $r \subseteq t$,

$\quad B.r$
$=\quad$ {definition of B }
$\quad (\underline{E} Q :: (\mathbf{A} f : f \epsilon r : Q \text{ in } f))$
$=\quad$ {definition of D , using $r = t \backslash (t \backslash r)$ }
$\quad (\underline{E} Q :: D.(t \backslash r)) \quad ,$

so that we can now reformulate demonstrandum (i) into

$$(\underline{E} Q :: D.(t \backslash t)) \Leftarrow (\mathbf{A} p : \#p = \#t - 1 : (\underline{E} Q :: D.(t \backslash p))) \quad ;$$

since $t \backslash t = \phi$ and by the one-to-one correspondence between figures of t and one-element tuples $t \backslash p$ with $\#p = \#t - 1$, this is equivalent to

(ii) $\quad (\underline{E} Q :: D.\phi)$ follows from $(\mathbf{A} f : f \epsilon t : (\underline{E} Q :: D.\{f\})) \quad .$

Note2 . The shape of the rightmost formula in (ii) —a simple quantification over figures rather than tuples— was the main reason why in our definition of D , t's complement entered the scene. We arrived at $D.r$ by the technique of replacing a constant by a variable, applied to $B.t$'s inner quantification $(\mathbf{A} f : f \epsilon t : Q \text{ in } f)$. Rather than replacing t , however, we replaced "invisible" constant ϕ —$t = t \backslash \phi$—, by variable r ; although both options were there, our choice was guided by

manipulative convenience.
End Note2 .

* *

Now that B has been removed from the demonstrandum, let us first consider some nice properties of D (for proofs see the appendix), viz.

(0) $[D.r \wedge D.s \equiv D.(r \cap s)]$
(1) $[D.r \Rightarrow D.s] \Leftarrow r \subseteq s$.

Property (0) gives us the slack to write $D.\phi$ as $D.r \wedge D.s$ for $r \cap s = \phi$, and, hence, to write target $(\underline{E}Q :: D.\phi)$ of (ii) as

(iiia)   find disjoint r and s such that
(iiib) $(Q : D.r)$ and $(Q : D.s)$ have a common solution   .

Property (1) enables us to exploit the antecedent of (ii) :   the latter gives us that each of the $(\#t)$ equations $(Q : D.\{f\})$ has a solution, and because from (1) we deduce $[D.\{f\} \Rightarrow D.s]$ for $f \in s$, we also have

(2)   for $s \subseteq t$, $(Q : D.s)$ has a bag of (at least) $\#s$ solutions, viz. for each f, $f \in s$, a solution of $(Q : D.\{f\})$.

So far, the properties of D —such as (2)— only admit satisfaction of (iiib) for *non*-disjoint r and s: $f \in r \cap s \wedge D.\{f\} \Rightarrow D.r \wedge D.s$; however, according to (iiia) we need disjoint r and s, so we need some more. We get it by exploiting the convexity of the figures of t as follows. Without proof we use

(3)   For a convex figure f and a bag of points X with convex hull XH , we have $(\underline{A}Q : Q \in X : Q \text{ in } f) \equiv (\underline{A}Q : Q \in XH : Q \text{ in } f)$,

with which we derive (see the appendix)

(4)   For convex hull XH of X and tuple r ,
    $(\underline{A}Q : Q \in X : D.r) \equiv (\underline{A}Q : Q \in XH : D.r)$,

a relation telling us that for a bag X of solutions of $(Q : D.r)$, X's convex hull consists of solutions as well. As a result, we can reformulate (iiib) as

(ivb)   with R and S bags of solutions of $(Q : D.r)$ and $(Q : D.s)$ respectively, i.e. with $(\underline{A} Q : Q \in R : D.r)$ and $(\underline{A} Q : Q \in S : D.s)$,

(ivc)   the convex hulls of R and S have a common point; all this for

(iva) r and s disjoint, viz. for (iiia) .

Condition (iva) can be met in various ways, e.g. any partitioning of t will do, and by (2) there is a one-to-one correspondence between such partitions of t and partitions satisfying (ivb) of the bag containing one solution of $(Q : D.\{f\})$ for each f in t . Therefore, we take for $R \cup S$ this bag of $\#t$ points and try to partition it into an R and S satisfying (ivc) ; if we are successful we are done.

Because nothing is known about the position of points, we can only use their number $\#t$ — $3 < \#t$ — and the yet unused planarity of the figures, i.e. we could do with a theorem like "A bag of > 3 points in the plane can be partitioned into two subbags whose convex hulls have a common point." . Note that the "3" and the "plane" are related: the statement does not hold in three-dimensional space. In view of Note0 about 3's irrelevance, we venture a generalization of the above statement and reinvent

Radon's theorem . In a real linear space of dimension $k - 1$, each bag of $> k$ points can be partitioned into two subbags whose convex hulls have a common point.
End Radon's theorem .

(For a proof see the appendix).

With Radon's theorem we have finished the step of the induction and, thereby, the whole proof. Taking Note1 into account, we now have proved the generalized theorem

Theorem . For t a tuple of convex figures in a space of dimension $k-1$, with $k \leq \#t$, we have: each k figures of t have a point in common $\Rightarrow$ all figures of t have a point in common.
End Theorem .

$$* \qquad *$$

We first present the appendix to fulfil the proof obligations postponed, and finally end with some remarks.

Appendix .

Proof of (0) . $\quad [D.r \wedge D.s \equiv D.(r \cap s)] \quad$.

$\qquad D.(r \cap s)$

$= \qquad$ {definition of D }

$\qquad (\underline{A} f : f \in t \backslash (r \cap s) : Q \text{ in } f)$

$= \qquad$ {domain split: $f \in t \backslash (r \cap s) \equiv f \in t \backslash r \vee f \in t \backslash s$}

$\qquad (\underline{A} f : f \in t \backslash r : Q \text{ in } f) \wedge (\underline{A} f : f \in t \backslash s : Q \text{ in } f)$

$= \qquad$ {definition of D }

$\qquad D.r \wedge D.s$

End Proof of (0) .

Proof of (1) . $\quad [D.r \Rightarrow D.s] \quad \Leftarrow \quad r \subseteq s \quad$.
From (0) with $r \cap s = r$ we derive $[D.r \wedge D.s \equiv D.r]$, i.e. $[D.r \Rightarrow D.s]$.
End Proof of (1) .

Proof of (4) . $\quad$ For XH the convex hull of X

$\qquad (\underline{A} Q : Q \in XH : D.r) \qquad\qquad\qquad (**)$

$= \qquad$ {definition of D }

$\qquad (\underline{A} Q : Q \in XH : (\underline{A} f : f \in t \backslash r : Q \text{ in } f))$

$= \qquad$ {interchange of quantifications}

$$(\underline{A}\,f : f \in t\backslash r : (\underline{A}\,Q : Q \in XH : Q \text{ in } f)) \qquad (*)$$
$$= \qquad \{(3)\}$$
$$(\underline{A}\,f : f \in t\backslash r : (\underline{A}\,Q : Q \in X : Q \text{ in } f))$$
$$= \qquad \{\text{undoing: } (*) \text{ and } (**) \text{ with } XH := X\}$$
$$(\underline{A}\,Q : Q \in X : D.r) \quad .$$

End Proof of (4) .

For completeness' sake we also include a proof of Radon's theorem.

Proof of Radon's theorem . We are to show that in a real linear space of dimension $k - 1$, each bag of $> k$ points can be partitioned into two subbags whose convex hulls have a common point. We define the convex hull of a bag of points to consist of all the linear combinations of the points such that each coefficient is ≥ 0 and the sum of the coefficients equals 1 .

We can now rephrase our interest in convex hulls with common point —x , say— into an interest in a linear combination, of all the points, that (a) is non-trivial, (b) is equal to 0 , i.e. to $x - x$, and (c) whose coefficients have sum 0 , i.e. $1 - 1$.

In a space of dimension $k - 1$, a linear combination of the more than k points that satisfies (b) gives rise to $k - 1$ homogeneous equations in the more than k coefficients, and property (c) gives another such equation. These equations have a non-trivial solution, and, by scaling, a non-trivial solution for which the sum of the positive coefficients equals 1 and hence —by (c)— that of the negative coefficients equals -1 .

From such a solution we construct a partition of the points into bags v and w such that the points with positive coefficients are in v and those with negative coefficients are in w . (The allocation of points with coefficient 0 is immaterial.) With Pv and Pw the linear combination confined to v and w respectively, we have $Pv + Pw = 0$, i.e. $Pv = -Pw$, while Pv is in v's convex hull and $-Pw$ is in w's convex hull.

End Proof of Radon's theorem .

Remark . In two respects our formulation of Radon's theorem is stronger than formulations found in the literature: it uses bags rather than sets of points, and "more than k points" rather than "exactly $k+1$ points". In this way the theorem provides a better interface: it smoothens application without complicating the proof.
End Remark .

End Appendix and Proof .

* *
*

The above proof, though as we later discovered not new, has been newly invented. The initial incentive to construct this formal calculational proof was a desire to avoid some of the characteristics of proofs we found in the literature: we wanted to avoid pictures and abundant nomenclature; in any case we wanted to avoid unmanipulatable formulae like

$$x_i \in F_1 \cap F_2 \cap \ldots \cap F_{i-1} \cap F_{i+1} \cap \ldots \cap F_r \neq 0 \quad \text{(cyclic)}$$

(see "Convex Sets" by Frederick A. Valentine). We achieved the latter by not naming all the figures and instead introducing the concept D, with which we can render the analogue of the above formula as $D.\{g\}$ for g a figure, and which invites us to formulate and use simple properties of D. (We mention that the above simple rendering $D.\{g\}$ is the main reason why t's complement entered the definition of D (see Note2).)

Besides having been included for its formal character, the present proof has been included for the way in which we chose to effectuate disentanglement in it. We were keen on being explicit about where the various data are used, or rather where they are not used, because we are interested in dealing with as few issues at the same time as possible.

The introduction of nomenclature did the job: by introducing B and C, we could remove C, and with it most traces of "3", before introducing more details of B by D and predicate Q in f; D enabled us to formulate a few simple heuristically helpful properties such as the exploitation of convexity in (4); and, finally, in predicate "Q in f" we encapsulated all geometric interpretations. (In passing we note that the need for such interpretations hardly arises: except for the validity of (3) and the adequacy of B's formalization in terms of D, the proof is independent of what Q in f stands for.) We note that as a byproduct of the encapsulation, Helly's theorem is independent of whether some of the figures coincide: coincidence of f and g can now only be expressed as $[Q$ in $f \equiv Q$ in $g]$, and such expressions do not occur in the argument.

In summary, the introduction of names, largely for predicates, has enabled us to keep formulae manageably short and, thereby, to give a formal proof, the shape of the formulae leading the way, and it has enabled the gradual introduction and exploitation of detail, disentangledness being a result.

10 The formal derivation of a proof of the invariance theorem

> The construction of the formal proof in this chapter illustrates
> to what extent the shape of formulae rather than their inter-
> pretation can inspire and assist the design of a proof.

In this chapter we formally derive a proof for the invariance theorem.
Our only concern here is the design of that proof from its specification:
we wish to show to what extent the symbols can do the work. For
the reader that has an interest in the theorem itself, we include a short
appendix. The contents of that appendix are, however, of no relevance
to the argument below.

Theorem . For

P and Q : predicates on a space V ;

t : an expression on V , having its values in a partially
            ordered universe D (" $<$ " denoting the non-reflexive
            partial ordering);

C : a subset of D ;

f : a predicate transformer;

we have that

$$[P \Rightarrow Q]$$

follows from the conjunction of

0. $(C, <)$ is well-founded
1. $[P \wedge \neg(t \in C) \Rightarrow Q]$
2. $(\underline{A} x : x \in D : [P \wedge t = x \Rightarrow f.(P \wedge t < x)])$
3. $[f.Q \equiv Q]$
4. f is monotonic   .

End Theorem .

Notational remark . Square brackets denote universal quantification over the space V ; t can be viewed as a function application in which both the function (from V to D) and the argument have been left implicit, while the whole expression has been named (similarly for P and Q).
End Notational remark .

We shall prove the theorem by proving $[P \Rightarrow Q] \Leftarrow$ **true** in a calculation starting at $[P \Rightarrow Q]$ and terminating at **true** , and using premisses (0) through (4). We note that t , C , and f —unlike P and Q — only occur in the premisses of the theorem and not in consequent $[P \Rightarrow Q]$ or in **true**. This means that somehow they have to enter the calculation and disappear from it again. The same holds for universal quantifier "$\underline{A}$" , in (2) .

We cannot say much about such introductions and removals until (0) and (4) are known in some more detail.

(4a) f is monotonic $\equiv$ for all Y and Z , $[f.Y \Rightarrow f.Z] \Leftarrow [Y \Rightarrow Z]$.

By its shape, definition (4a) looks most suitable for removing f 's from the calculation of $[P \Rightarrow Q] \Leftarrow$ **true** .   Premiss (3) — $[f.Q \equiv Q]$ — seems suitable for both introduction and removal.

(0a) $(C, <)$ is well-founded

$\equiv$

for each predicate S

$[(\underline{A} x : x \in C : S.x) \equiv (\underline{A} x : x \in C : S.x \Leftarrow (\underline{A} y : y \in C \wedge y < x : S.y))]$.

The well-foundedness does not help us introduce or remove $\underline{A}$'s, it allows us to rewrite one quantification equivalently into a formally weaker one.

Since we do not have a rule for introducing "$\underline{A}$" yet, it may be worthwhile to remember the one-point rule:

(5) $\quad [E(x := r) \equiv (\underline{A}\,x : x = r : E)]$ for any expressions r and E .

We have one final remark before starting the calculation. In the derivation, there are only a few places where a design decision has to be taken. We give the calculation first and discuss the design decisions later. For ease of reference, we label the steps that embody these decisions.

*

Proof .

(6) $\quad$ Instead of massaging $[P \Rightarrow Q]$, we massage the more general $[P \wedge Z.t \Rightarrow Q]$, for $Z.t$ a predicate:

$$[P \wedge Z.t \Rightarrow Q]$$
(7) $=\quad$ $\{\,(1)\,,$ i.e. $[Q \equiv Q \vee (P \wedge \neg(t \in C))]\,\}$
$$[P \wedge Z.t \Rightarrow Q \vee (P \wedge \neg(t \in C))]$$
$=\quad$ $\{$reshuffling, aiming at the removal of one $P\,\}$
$$[P \wedge Z.t \wedge \neg(P \wedge \neg(t \in C)) \Rightarrow Q]$$
$=\quad$ $\{$De Morgan, negation$\}$
$$[P \wedge Z.t \wedge (\neg P \vee t \in C) \Rightarrow Q]$$
$=\quad$ $\{$complement rule$\}$
$$[P \wedge Z.t \wedge t \in C \Rightarrow Q]$$
(8) $=\quad$ $\{\,(5)$ with $r := t$, $E := P \wedge Z.x \wedge x \in C \Rightarrow Q\,;$
$\quad\quad$ see Note0 below$\}$
$$[(\underline{A}\,x : x = t : P \wedge Z.x \wedge x \in C \Rightarrow Q)]$$
(9) $=\quad$ $\{$trading, heading for (2)'s antecedent$\}$

$$[(\underline{A}\, x\, :\, Z.x \wedge x \in C : P \wedge x = t \;\Rightarrow\; Q)]$$

$\Leftarrow$ {see Intermezzo below, manipulating the term,

monotonicity of $\underline{A}$ and $[\,]\,\}$

$$[(\underline{A}\, x\, :\, Z.x \wedge x \in C : [P \wedge t < x \;\Rightarrow\; Q])] \qquad .$$

Note0 . Here we use that $[P(x := t) \equiv P]$, similarly $[Q(x := t) \equiv Q]$,
and $[(Z.x)(x := t) \equiv Z.t]$.
End Note0 .

Deferring the Intermezzo, which exploits (2), (3), and (4)/(4a),
to the end of the proof, we summarize, noting that the only unused
premiss is (0)/(0a), that our calculation so far yields

(10) $[P \wedge Z.t \;\Rightarrow\; Q] \;\Leftarrow\; [(\underline{A}\, x\, :\, Z.x \wedge x \in C : [P \wedge t < x \;\Rightarrow\; Q])]$.

Note1 . In (10), the square brackets around $(\underline{A}\, x\, :\, Z.x \ldots)$ may be
omitted if $Z.x$ is independent of V.
End Note1 .

Now we can perform the main calculation, exploiting (0), by

$$[P \Rightarrow Q]$$

$\Leftarrow$ { (10) with $Z.t :=$ **true** }

$$[(\underline{A}\, x\, :\, x \in C : [P \wedge t < x \;\Rightarrow\; Q])]$$

$=$ { (0)/(0a) with $S.x := [P \wedge t < x \;\Rightarrow\; Q]$ }

$$[(\underline{A}\, x\, :\, x \in C : [P \wedge t < x \;\Rightarrow\; Q] \;\Leftarrow$$
$$(\underline{A}\, y\, :\, y \in C \wedge y < x : [P \wedge t < y \;\Rightarrow\; Q]))]$$

$=$ { (10) with $Z.t := t < x$, $x := y$, see Note1}

$$[(\underline{A}\, x\, :\, x \in C : \textbf{true})]$$

$=$ {predicate calculus}

$$[\ \textbf{true}\ ]$$

$=$ { }

true   ,

thus fulfilling our main proof obligation.

Finally we have

Intermezzo . For $x \in C \wedge Z.x$

$$P \wedge x = t \Rightarrow Q$$
$\Leftarrow \qquad$ {heading for (2)'s consequent, transitivity of $\Rightarrow$ }
$$(P \wedge x = t \Rightarrow f.(P \wedge t < x)) \wedge (f.(P \wedge t < x) \Rightarrow Q)$$
$= \qquad$ { (2) using $C \subseteq D$ }
$$f.(P \wedge t < x) \Rightarrow Q$$
$= \qquad$ { (3) , preparing removal of f's by (4)/(4a) }
$$f.(P \wedge t < x) \Rightarrow f.Q$$
$\Leftarrow \qquad$ { (4)/(4a) }
$$[P \wedge t < x \Rightarrow Q] \qquad .$$

End Intermezzo .

End Proof .

*

On the design decisions .

Our first design decision was to try to use each of the premisses of the theorem only once, because in our experience duplicate usage can be a symptom of insufficient disentanglement of the argument. Here the decision led us to the isolation of lemma (10) , i.e. to design decision (6): before we had decided to massage the more general $[P \wedge Z.t \Rightarrow Q]$ instead of $[P \Rightarrow Q]$, our calculation more or less consisted in doing the same manipulations twice.

So design decision (6) helps avoid repetition and lengthiness, but the price we pay is that now the reader is confronted with this decision at a stage of the calculation at which he cannot yet see the need for it. In general we prefer to avoid such heuristic rabbits being pulled out of a hat as much as possible, but as illustrated here, we do not do so at any cost.

As for the design decision embodied in the first step, (7), of the calculation: few of the options look promising; premisses (0), (2), and (4) are not electable since there is no "f" or "$\underline{A}$" in manipulandum $[P \wedge Z.t \Rightarrow Q]$, and introducing one f by (3) does not help much yet. What remains is the choice between the use of (1) and the introduction of an "$\underline{A}$" by (5).

Of the two options, the introduction of "$\underline{A}$" is, however, some-what premature because we do not know what to choose for r in (5). Hence we first exploit premiss (1) in step (7), simplify, and then are ready in step (8) to introduce the universal quantifier.

Having introduced the "$\underline{A}$", we now have two options: using (0)/(0a) or first exploiting (2). The first choice would give us a long expression to which (2), (3), and (4) still have to be applied, so for the sake of brevity and simplicity we choose —in step (9)— to head for the exploitation of (2) first. After this decision, the rest of the calculation hardly leaves any more choice.

We note that the emergence of expression $[P \wedge t < x \Rightarrow Q]$ at the end of the intermezzo, where all premisses except (0) have al-ready been used, is a strong hint that $[P \wedge Z.t \Rightarrow Q]$ could be worth considering. Indeed, its emergence in our original repetitious calcula-tion inspired the transition to manipulating this more general expression instead of the original demonstrandum $[P \Rightarrow Q]$.

*

In summary, the design given above rests on three pillars: syntactic analysis of the formulae to be manipulated, familiarity with predicate calculus so as to be able to mould the formulae into a shape that is convenient for the manipulations to be performed, and the use of a few of our rules of thumb, such as trying to use each datum only once and trying to postpone usage until hardly anything else can be done.

Showing these three things at work was our main goal. We had proved our vehicle towards that goal —the invariance theorem— before, e.g. in "A simple fixpoint argument without the restriction to continuity", Edsger W. Dijkstra and A.J.M. van Gasteren, in Acta Informatica. It took us quite a few iterations to get the proof presented in that paper in a sufficiently nice form; the design given here emerged with W.H.J. Feijen's cooperation, after he had suggested that in the mean time we should have learned to *construct* a proof inspired by the shape of the formulae.

*

Appendix .

In the usual invariance theorem for the repetition, P is the invariant and t the variant function of repetition DO: **do** $B \rightarrow S$ **od** , whose semantics is considered equivalent to the semantics of
if $B \rightarrow S$; DO $[] \neg B \rightarrow$ **skip fi** , viz. for all R

$$[\text{wp}.(\text{DO}, R) \equiv (B \vee R) \wedge (\neg B \vee \text{wp}.(S, \text{wp}.(\text{DO}, R)))] \quad .$$

For $R := P \wedge \neg B$ in particular, we have that $\text{wp}.(\text{DO}, P \wedge \neg B)$ is a solution of $(X : [X \equiv f.X])$, with $[f.X \equiv (B \vee P) \wedge (\neg B \vee \text{wp}.(S, X))]$. (It is defined to be the strongest solution.)

That is what premiss (3) comes from. Premiss (4), f's monotonicity, is usually formulated as $\text{wp}.(S, ?)$ is monotonic. Premiss (2) is the usual $(\underline{\text{A}} x :: [P \wedge B \wedge t = x \Rightarrow \text{wp}.(S, P \wedge t < x)])$, and premiss (1) is a (weakened) substitute for the traditional $[P \wedge B \Rightarrow t \in C]$. The latter two statements require a proof, which we leave to the reader. (Predicate calculus suffices.)

End Appendix .

11 Proving theorems with Euclid's algorithm

Algorithms can be used to prove theorems. This chapter illustrates how the notion of invariance can assist in proving equivalences directly instead of by mutual implication.

Most monographs on number theory include Euclid's algorithm, in some form or other, in their treatment of the greatest common divisor. Usually, the validity of the procedure rests in some way or other, on the validity, for all integers z, v, and w, of

(0) $\quad z|v \wedge z|w \equiv z|v \wedge z|(w - v)$

(1) $\quad z > 0 \Rightarrow z \operatorname{gcd} z = z \quad .$

Rarely, however, is the algorithm used at all. In this small chapter we shall show how it can be used to prove some theorems. In the following, variables denote integers unless stated otherwise.

*   *
*

We use the algorithm for computing $p \operatorname{gcd} q$ in the following form

$$
\begin{aligned}
&|[\ p, q : \text{int } \{p > 0 \ \wedge \ q > 0\} \\
&;\ |[\ x, y : \text{int}\,;\ x, y := p, q \\
&\quad ;\ \mathbf{do}\ x > y\ \rightarrow\ x := x - y \\
&\qquad\quad []\ y > x\ \rightarrow\ y := y - x
\end{aligned}
$$

od
  $\{x = y\}$
 ]|
]| .

Using (0), the reader may verify that the above program maintains P,

P: $\quad (\underline{\mathbf{A}}\, z :: z|p \wedge z|q \equiv z|x \wedge z|y) \wedge x > 0 \wedge y > 0$,

and, hence by the notion of "greatest",

$$p \gcd q = x \gcd y \wedge x > 0 \wedge y > 0 \quad ,$$

so that upon termination, from $x = y$ and (1), we have $p \gcd q = x$ $\wedge\ p \gcd q = y$.

All this is familiar to computing scientists, but perhaps a little less so to other mathematicians. The reason why we exhibit it is to stress how the whole almost solely depends on the simple additive property (0), and to pave the way for the observation that if instead of exploiting (0) we exploit the equally valid (2):

(2) $\quad z|m*v \wedge z|m*w \equiv z|m*v \wedge z|m*(w - v)$,

we similarly have the invariance of Q:

Q: $\quad (\underline{\mathbf{A}}\, z :: z|m*p \wedge z|m*q \equiv z|m*x \wedge z|m*y)$;

then, upon termination, with x and y equal to $p \gcd q$, we hence have

(3) $\quad (\underline{\mathbf{A}}\, z :: z|m*p \wedge z|m*q \equiv z|m*(p \gcd q))$, for $p > 0 \wedge q > 0$.

Since (3) does not depend on variables of the state space, it not only holds upon termination, but also is just a valid theorem.

We note that (3) also holds for non-positive integers p and q. It holds for $p < 0$, since

$$(-p) \gcd q = p \gcd q \quad \text{and} \quad z|m*(-p) \equiv z|m*p \quad ;$$

it holds for $p = 0$ and $q \neq 0$, since

$$0 \gcd q = q \gcd q \quad \text{and} \quad z|0 \wedge z|m*q \equiv z|m*q \wedge z|m*q \quad .$$

Hence using the symmetry between p and q we have

(4) $(\underline{A} z :: z|m*p \wedge z|m*q \equiv z|m*(p \gcd q))$, for $p \neq 0 \vee q \neq 0$.

Different instantiations of (4) now immediately provide a number of well-known theorems about greatest common divisors:

Corollary0 . Instantiation of (4) with $m = 1$ yields the familiar

$$(\underline{A} z :: z|p \wedge z|q \equiv z|p \gcd q) \quad .$$

Corollary1 . Instantiation of (4) with $z = p$ yields

$$p|m*q \equiv p|m*(p \gcd q) \quad ,$$

which is probably less familiar than its consequence

Corollary1a . If $p \gcd q = 1$ and $p|m*q$ then $p|m$.

Furthermore we have, for any z ,

$$z|m*(p \gcd q)$$
$$= \qquad \{(4)\}$$
$$z|m*p \wedge z|m*q$$
$$= \qquad \{\text{Corollary0}\}$$
$$z|(m*p) \gcd(m*q) \quad ,$$

from which we conclude —using that $(\underline{A} z :: z|x \equiv z|y) \equiv |x| = |y|$ and that gcd is positive—

Corollary2 . $(m*p) \gcd(m*q) = |m|*(p \gcd q)$, i.e. multiplication with a positive number distributes over gcd .

Corollary3 . gcd is as associative as conjunction: we have, for any z ,

$$z|(p \gcd q) \gcd r$$
$$= \qquad \{\text{Corollary0}\}$$
$$z|(p \gcd q) \wedge z|r$$
$$= \qquad \{\text{Corollary0}\}$$
$$(z|p \wedge z|q) \wedge z|r \quad , \qquad\qquad \text{and similarly}$$

$$z|p \wedge (z|q \wedge z|r) \equiv z|p \gcd(q \gcd r) \quad ,$$

so that

conjunction is associative
$$\Rightarrow \qquad \{\text{see above}\}$$
$$(\underline{\mathbf{A}} z :: z|(p \gcd q) \gcd r \equiv z|p \gcd(q \gcd r))$$
$$= \qquad \{\text{arithmetic, and gcd is positive}\}$$
$$(p \gcd q) \gcd r = p \gcd(q \gcd r) \quad .$$

$$* \qquad\qquad *$$
$$*$$

The above theorems can be, and are, proved in various ways. We have several reasons for showing our proofs, a main one being that the algorithm is such an effective interface: the proofs of the corollaries require no appeal whatsoever to properties of gcd ; all such appeals are confined to the proof that Euclid's algorithm meets its requirements. We consider this a nice separation of concerns. The most elegant "conventional" proof we have seen for, for instance, Corollary1a, requires the introduction of two extra names —as our algorithm does— but relies on the theorem that $p \gcd q$ is a linear combination of p and q (a theorem that can also be proved with the algorithm, viz. by choosing "x and y are linear combinations of p and q" as an additional invariant). Other proofs rely on the unique prime factorization property.

A second reason for showing the above is that the equivalence in (4) was not proved by proving mutual implication, in contrast with traditional proofs of the corollaries, which usually are.   In fact, the validity of (0) and (2) together with the notion of invariance is a strong incentive to consider such equivalences as appearing in P and Q.

We also note that by choosing to use formula (4) we opened the way for such simple conclusions as Corollary3, viz. that gcd is as associative as conjunction, and achieved that further proof obligations could largely be met by manipulation of uninterpreted formulae.

12 On the design of an in-situ permutation algorithm

It is shown how the availability of an adequate notation, for permutation-valued expressions in this case, can be essential for the presentation of an algorithm and the design decisions leading to it. The choice of the notation was guided by constraints of manipulability, constraints that were met primarily by being frugal in the use of nomenclature (of subscripted variables in particular).

In this chapter we develop an algorithm for the in-situ inversion of a cyclic permutation that is represented in an array. We do so not for the sake of the algorithm, but because the development is such a clear demonstration of how the availability of an adequate notation can decide between failure and success.

We present the development of the algorithm first. In a "heuristic" epilogue we shall discuss some of the design decisions in more detail. In particular, we shall pay attention to the choice of our notation and to the reasons why alternative notations that we employed in earlier efforts were inadequate.

*   *
*

We consider a permutation P of the elements of a finite non-empty universe, i.e. P is a one-to-one function from the universe to the universe. Its inverse Q is defined by

(0) $Q.j = i \equiv P.i = j$,   for each i, j in the universe.

We want to design an algorithm S that changes in situ an array H initially being equal to a cyclic permutation P into the array representing its inverse Q; i.e. the functional specification of S is

(1) $\{H = P\}\ S\ \{H = Q\}$,   for P and Q satisfying (0).

*

Besides introducing some notation and nomenclature, we first introduce the concept of a "ring" and its relation with cyclic permutations and arrays. After having done so, we shall carry out the major part of the development in terms of such rings, because they are more conveniently manipulated than arrays. (Also, the treatment will pertain to permutations over any universe, i.e. by choosing some one-to-one correspondence between such a universe and an initial segment of the natural numbers, the ultimate algorithm may be used for arbitrary domains.)

•   Elements of the universe are denoted by lower case letters, sequences of such elements by capitals, and catenation of finite sequences by juxtaposition; one-element sequences and elements are identified; the empty sequence is denoted by ϕ.

•   Function rev on finite sequences is defined by

$$
\begin{aligned}
rev.\phi   &= \phi \\
rev.d   &= d \\
rev.(dC)   &= (rev.C)\,d   &&,\quad \text{hence satisfies} \\
rev.(BC)   &= (rev.C)(rev.B)   &&.
\end{aligned}
$$

•   On finite sequences of distinct elements we define equivalence classes, called "rings", denoted with square brackets and induced by the equivalence relation relating BC to CB, for all finite sequences B and C; i.e. we have

$[BC] = [CB]$,   the Rule of Rotation.

● For elements in a non-empty ring the notion of "follower" is defined by

(2)   the follower of d in $[BdC]$ = the first element of sequence CBd .

(Note that in a one-element ring, the element is its own follower.)

● Finally, between cyclic permutations and rings we establish a one-to-one correspondence: for cyclic permutation R and ring $[B]$ that have the same domain

(3) R corresponds to $[B]$ $\equiv$ $(\mathbf{A} i :: $ the follower of i in $[B]$ is $R.i)$.

The importance of this correspondence is that for P and Q satisfying (0) we have

(4) P corresponds to $[U]$ $\equiv$ Q corresponds to $[rev.U]$.

The (calculational) proof of (4) is given in an appendix.
End ● .

Correspondence (3) and theorem (4) enable us to rephrase specification (1) in terms of rings as

$$\{h = [U]\}\ S\ \{h = [rev.U]\}$$

where h is a variable of type ring (corresponding to the original H) and $[U]$ is a constant of type ring (corresponding to P).

The definition of rev shows that for rings with at most one element, $S = $ skip satisfies the specification;   therefore, we concentrate on rings with at least two elements, viz. we solve, with p and q as two further constants, of type element,

(5) $\{h = [Upq]\}\ S\ \{h = [qp\,(rev.U)]\}$.

Remark0 .   The above case analysis requires a justification.   We deal with that in the epilogue.
End Remark0 .

*

In order to find an invariant, we consider generalizations of the initial and final states in (5). Observing that, by the Rule of Rotation, the postcondition is equivalent to $h = [p\,(rev.U)\,q]$, we are led to the introduction of two new variables X and Y , of type sequence, and connected with h by invariant

P0 $h = [XpYq]$.

Solving equation $(X, Y : P0)$ for initial and final state yields

(6)   initially, $X, Y = U, \phi$
           finally, $X, Y = \phi, rev.U$.

For the time being we postpone our concerns about variable h and concentrate on finding a program fragment operating on the sequence variables X and Y and establishing transformation (6). Since initially and finally the sequences X and Y comprise the elements of U , it is strongly suggested to include this in an invariant. We suggest

P1 $(rev.Y)\,X = U$,

which is satisfied by both states of (6) —on account of properties of rev — .

Remark1 . Why we did not propose the at this stage equally acceptable invariant $X(rev.Y) = U$ will be discussed in the epilogue.
End Remark1 .

Because $P1 \wedge X = \phi$ implies the final state, the task is to shrink X under invariance of $P1$. Doing so one element at the time means —see $P1$ — removing the first element of X and attaching it to the end of $rev.Y$, i.e. to the front of Y .

So we get for transition (6) the program fragment

        do $X \neq \phi \rightarrow$
                    with r and Z chosen to satisfy $X = rZ$:
                    ; $X, Y := Z, rY$
        od   .

So much for the invariance of $P1$.

It is now time to return to h and the invariance of $P0$. For the latter purpose we introduce the statement "massage h", whose refinement will be postponed for a while. We insert it before "$X, Y := Z, rY$", so as to be able to derive its postcondition with the axiom of assignment. Including the initialization, we get our next version.

$$X, Y := U, \phi \; \{P0 \wedge P1\}$$
$$; \textbf{do } X \neq \phi \; \rightarrow$$
$$\qquad \text{with } r \text{ and } Z \text{ chosen to satisfy } X = rZ :$$
$$\qquad ; \text{"massage } h\text{"}$$
$$\qquad ; X, Y := Z, rY \; \{P0 \wedge P1\}$$
$$\textbf{od}$$
$$\{P0 \wedge P1 \wedge x = \phi\}$$

The only task left is the design of "massage h" so as to guarantee the invariance of $P0$. We derive a specification for "massage h" by taking $P0 \wedge X = rZ$ as its precondition and wp.$(X, Y := Z, rY, \; P0)$ as its postcondition:

(7) $\qquad \{h = [rZpYq]\}$ massage h $\{h = [ZprYq]\}$.

We shall translate operations of "massage h" into operations on H according to

(8) $\qquad H$ corresponds to ring h

Finally, h, U, X, Y, and Z will be eliminated as thought quantities.

*

Starting with the refinement of "massage h", we note that (8) together with (7)'s precondition gives:

$$H.p \; = \; \text{the first element of sequence } Yq$$
$$H.q \; = \; r$$
$$H.r \; = \; \text{the first element of sequence } Zp \quad ,$$

and that (8) together with (7)'s postcondition gives:

$$H.p = r$$
$$H.q = \text{the first element of sequence } Zp$$
$$H.r = \text{the first element of sequence } Yq \quad .$$

For all other elements of ring h (viz. the elements of Z and Y), the follower in h does not change. Hence, "massage h" can be refined to

$$\text{"massage } h\text{"} \; : \; H.p, H.q, H.r := H.q, H.r, H.p \quad .$$

Note that here we use that the elements of a ring are distinct.

Finally we observe, using correspondence (8), that

- in the initial state of S (see (5)), $q = H.p$;
- by $P0$, guard $X \neq \phi$ is equivalent to $H.q \neq p$;
- by $P0$ and $X = rZ$, $r = H.q$.

Hence, thought variables h, U, X, Y, and Z can be eliminated, yielding the ultimate program

$$\{p \text{ is any element of the ring to be inverted }\}$$
$$q := H.p$$
$$; \textbf{do } H.q \neq p \rightarrow$$
$$r := H.q$$
$$; H.p, H.q, H.r := H.q, H.r, H.p$$
$$\textbf{od} \quad .$$

(**Note** . If so desired, the above program can be changed a little so as to maintain $r = H.q$; in that way the duplication of expression $H.q$ can be avoided. Bothering about such issues, however, is beyond the scope of this text.
End Note .)

$$*\qquad\qquad*$$
$$*$$

Before discussing other matters, we first return to the design decisions referred to in Remark0 and Remark1.

Re Remark0 . We could justify our distinction between rings with at most one element and rings with at least two by pointing out that the program we derived works for rings with one element also, but that would be too easy a way out: it is a justification after the fact rather than a well-considered design decision.

We do, however, have a more convincing justification. We did *not* start our initial investigation with the case distinction, because as a rule we avoid making distinctions unless we cannot. So from the specification, $\{h = [U]\}\ S\ \{h = [rev.U]\}$ we derived —by the same pattern of reasoning as used in the development above— invariants

$Q0 \qquad h = [XY]$
$Q1 \qquad [U] = [(rev.Y)X] \quad , \qquad$ with

initially, $\quad U = XY \wedge Y = rev.Y \quad$ and
finally, $\quad\ X = rev.X \quad$.

The same shrinking of X , by $\{X = rZ\}\ X, Y := Z, rY$, now yields for the specification of "massage h" , (see $Q0$),

$$\{h = [rZY]\}\ \text{massage}\ h\ \{h = [ZrY]\} \quad .$$

For the implementation of this "massage h" in terms of array H , we then noted that $H.r$, i.e. the follower of r in h , changes only if neither Z nor Y is empty.

So we were in a quandary: admitting the case $Z = \phi \vee Y = \phi$ would give us a case analysis in the repeatable statement, excluding it by adding $Z \neq \phi \wedge Y \neq \phi$ as a precondition of the repeatable statement would force us to confine the algorithm to rings with at least two elements, i.e. it would give us a case analysis "outside the repetition". If forced to choose between the two, however, we as a rule prefer the latter. Hence our choice.

End Re Remark0 .

Re Remark1 .   Concerning the choice between $X(rev.Y) = U$ and $(rev.Y)X = U$ as an invariant, we note two things. Firstly, whatever the choice, it is uniquely connected to the choice between shrinking X at the end or shrinking it at the front: the one choice fixes the other. Secondly, the choice of how to shrink X is fixed by the chosen correspondence between ring h and array H and our wish to design an efficient algorithm: followers can be computed efficiently, viz. by one application of H , other elements cannot. Given $h = [XpYq]$, the computation of X's last element takes as many applications of H as X has elements. **End** Re Remark1 .

*   *

*

We included the above justification of our earlier design decisions for several reasons. We did so firstly because we like the degree of completeness that can be achieved for such considerations. Secondly, we consider the analysis yet another example of how a careful syntactic analysis of the formulae gives strong heuristic guidance. But most importantly, we wanted to show how vital an adequate notation is for the feasibility of such an analysis. We are convinced that the above analysis could not have been carried out without the introduction of the concept of a ring and the Rule of Rotation.

Because the introduction of the notion ring is at the heart of the design, we give in short the history of its invention. In two earlier efforts at developing a formal and convincing argument for the permutation inversion problem, we had not introduced the concept yet. The treatments were, indeed, formal but barely convincing. It was only later, when we had experienced more often how tightly the choice of notation is coupled to the manipulations required, that we realized what the source of the trouble had been.

As soon as we started to develop a little calculus for permutations, it became clear that in order to keep the formulae manageable, we should introduce only a very modest amount of nomenclature. And that was exactly where the earlier efforts had failed miserably.

In the first effort we characterized a cyclic permutation in terms of its array representation H by giving one element p and giving the others by expressions of the form $H^k.p$. Such expressions then occurred in large quantities in the text, quantified and not quantified, making first of all the formulae almost unmanageable and making, secondly, the argument totally unconvincing, because the effect of a change of array H on expression $H^k.p$ could not be described simply.

The next effort was superior in that it at least introduced an auxiliary concept, but its relation with cyclic permutations was as inadequate as the above one: the concept was a sequence of appropriate length with some rotational freedom, and the relation between sequence s and permutation y was given by $s.i = y^i.m$ for all i in the given range and some fixed m. As a result the treatment, besides suffering from indexitis, was as much complicated by the occurrence of expressions $y^i.m$ as the first effort.

In both efforts the major mistake was that all elements, and even the size, of the permutation had received a name, overwhelming detail and unwieldy formulae being the result. With such overspecific nomenclature, even the formulation of a simple rule like the Rule of Rotation would have been painful, and any discussion of heuristic considerations nearly impossible. Although conventional, the notations for permutations that we had used, appeared to be totally inadequate for our manipulative needs.

That our third effort became an order of magnitude more effective than the earlier ones, again confirms our and many computing scientists' opinion that in the design of algorithms the development of adequate mathematical notations is a key issue.

Finally we mention that our initial incentive to investigate the permutation problem dealt with here was the publication of an algorithm for it in Information Processing Letters by B.-C. Huang. Because that paper did not give any justification for the correctness of the solution, we decided to investigate the problem in order to find out how such permutation problems could be dealt with more formally and convincingly. We confined ourselves to cyclic permutations because that is where the heart of the problem is.

The above treatment is a revised and extended version of a paper co-authored by W.H.J. Feijen and D. Gries, published in Information Processing Letters.

Appendix . Proof of (4): for P and Q satisfying (0)

P corresponds to $[U] \equiv Q$ corresponds to $[rev.U]$.

The validity of (4) rests on the validity of two relations between rev and rings:

(9) $[X] = [Y] \Rightarrow [rev.X] = [rev.Y]$ and

(10)   the follower of i in $[U]$ is $j \equiv$ the follower of j in $[rev.U]$ is i

which will be dealt with shortly.

P corresponds to $[U]$

=   {definition (3) of "corresponds"}

($\underline{A} i ::$ the follower of i in $[U]$ is $P.i$)

=   { (10) }

($\underline{A} i ::$ the follower of $P.i$ in $[rev.U]$ is i)

=   {dummy transformation: $i = Q.j$, i.e. by (0) , $P.i = j$}

($\underline{A} j ::$ the follower of j in $[rev.U]$ is $Q.j$)

=   {definition (3)}

Q corresponds to $[rev.U]$.

Relation (9) holds since for all B and C

$$[rev.(BC)] = [rev.(CB)]$$
$=$ {definition of rev}
$$[(rev.C)(rev.B)] = [(rev.B)(rev.C)]$$
$=$ {Rule of Rotation}
true ;

relation (10) holds since for $U = XiY$

j is follower of i in $[U]$
$=$ {definition (2) of follower, $U = XiY$}
j is the first element of YXi and i is last
$=$ {property of rev}
j is last element of $rev.(YXi)$ and i is first
$=$ {definition of follower}
i is follower of j in $[rev.(YXi)]$
$=$ {by (9) $[rev.(YXi)] = [rev.(XiY)]$, $U = XiY$}
i is follower of j in $[rev.U]$.

End Appendix .

13 Shiloach's algorithm

> This chapter's purpose is to show how the use of an adequate
> formalism, predicate calculus in this case, enables us to present
> an algorithm clearly, concisely, and in all relevant detail, in a
> way that reveals all the ingenuities of the design.

As an illustration of how a program can be presented clearly, concisely,
and in all its relevant detail we present Shiloach's Algorithm for checking
the equivalence of two circular lists. The presentation is followed by a
discussion.

The problem we describe is symmetric in the integer sequences
$A(i : 0 \leq i < N)$ and $B(i : 0 \leq i < N)$, $N \geq 1$, and so will be the solution.

In terms of $A(i : 0 \leq i < N)$ we define the set of so-called A-
sequences $SA.i$ —the "rotations" of sequence A— for all natural i
(i.e. $i \geq 0$) by

$$SA.i = A(k : i \leq k < i + N) \quad ,$$

with indices in A reduced modulo N. Note that $SA.i = SA.(i+N)$, so
that the set of A-sequences contains at most N elements. B-sequences
are defined by symmetry. We are requested to write a program solving
equation R in eq

$R: \quad eq \equiv (\underline{E}i, j :: SA.i = SB.j) \quad .$

Note . In the notation of Chapter 12: we are to solve equation

80

$$eq \equiv [A] = [B] \quad .$$

End Note .

Remark on terminology

In discussing programs, it is convenient not to restrict oneself to deterministic programs or program components. In order to do justice to this fact, we prefer to avoid terminology that is only applicable in the case of determinacy; in particular, we do not describe the "output" as a function of the "input". Instead, we specify the program using a postcondition to be satisfied at program completion. Such a postcondition is a boolean expression on what are also known as "input variables" and "output variables", the distinction being that the postcondition has to be satisfied by assigning suitable values to the output variables.

We have adopted the convention of rendering this distinction by considering the postcondition as an equation in the output variables and, accordingly, phrase the programmer's task as writing a program "solving that equation". (Here the English word "equation" has been generalized to that of the Dutch "vergelijking", in the sense that the postcondition may be any boolean expression and need not be constrained to equalities between two expressions.)

End Remark on terminology .

$$*$$

R is easily solved by comparing each A-sequence with each B-sequence, but in our trade we refuse to do so: no specific property of A-sequences and B-sequences would then be exploited.

We observe that the sets of A- and B-sequences are either disjoint, in which case $eq \equiv$ **false** solves R , or not disjoint, in which case $eq \equiv$ **true** solves R . If they are not disjoint they are equal: if $SA.i = SB.j$ then $SA.(i + k) = SB.(j + k)$, for all k , i , and j . The sets being either disjoint or equal, they can be compared by comparing canonical members, e.g. with

$$AA = \text{the lexicographically first } A\text{-sequence} \quad , \text{ and}$$
$$BB = \text{the lexicographically first } B\text{-sequence}$$

we have

$$(0) \qquad R \;\equiv\; (eq \equiv (AA = BB)) \qquad .$$

$$*$$

R can be solved by computing AA and BB, but in our trade we refuse to do so: in solving R only their equality or difference matters.

We propose to discover the solution $eq \equiv \textbf{true}$ by identifying a pair (i,j) such that $SA.i = SB.j$. To that end we introduce the weaker condition

$$P: \qquad 0 \leq h \;\wedge\; (\underline{A}\,k : 0 \leq k < h : SA.i.k = SB.j.k) \qquad ,$$

which derives its importance from the validity of

$$(1) \qquad (P \;\wedge\; h \geq N) \;\Rightarrow\; \text{wp}.(eq := \textbf{true}, R) \qquad :$$

$$P \;\wedge\; h \geq N$$
$$\Rightarrow \qquad \{\text{definition of } SA.i \text{ and } SB.j\,\}$$
$$SA.i = SB.j$$
$$= \qquad \{\text{definition of } R\,\}$$
$$\text{wp}.(eq := \textbf{true}, R) \qquad .$$

We propose to discover the solution $eq \equiv \textbf{false}$ by observing $AA \neq BB$. To that purpose we introduce the weaker condition

$$QA: \qquad 0 \leq i \;\wedge\; (\underline{A}\,k : 0 \leq k < i : SA.k > BB) \qquad ,$$

in which " $>$ " is read "comes lexicographically after". It derives its importance from the validity of

(2) $(QA \wedge i \geq N) \Rightarrow \text{wp.}(eq := \textbf{false}, R)$:

$(QA \wedge i \geq N)$

$\Rightarrow$ {definition of QA }

$(\underline{A} k : 0 \leq k < N : SA.k > BB)$

$=$ {definition of AA }

$AA > BB$

$\Rightarrow$ {see (0) }

$\text{wp.}(eq := \textbf{false}, R)$.

Note that condition P couples i with A and j with B, so that we propose that

(3)   the discussion is symmetric in the pairs (A, i) and (B, j) ,

from which QB and its properties follow.

The above observations suggest the following program:

```
|[ h, i, j : int
; h, i, j := 0, 0, 0
   {P ∧ QA ∧ QB}
; do h < N ∧ i < N ∧ j < N   →
               {h + i + j  ≤  3*N−3}
               "increase h + i + j maintaining P ∧ QA ∧ QB"
  od
   {P ∧ QA ∧ QB ∧ (h ≥ N ∨ i ≥ N ∨ j ≥ N)}
; if  h ≥ N   →   {P ∧ h ≥ N}  eq := true  {R, see (1)}
  []  i ≥ N   →   {QA ∧ i ≥ N} eq := false {R, see (2)}
  []  j ≥ N   →   eq := false {R, by (3) and the preceding
                                  alternative}
   fi {R}
]|  .
```

Note that the repetitive construct terminates by virtue of its specification.

Remark . The if-statement can be replaced by the much shorter $eq :=$ $(h \geq N)$ since its precondition guarantees that $i \geq N \lor j \geq N \Rightarrow$ $\neg(h \geq N)$:

$$QA \land i \geq N \land P$$

$\Rightarrow$ {see (2)}

$$AA \neq BB \land P$$

$=$ {see (0) , definition of P}

$$(\underline{A} i, j :: SA.i \neq SB.j) \land 0 \leq h \land$$
$$(\underline{A} k : 0 \leq k < h : SA.i.k = SB.j.k)$$

$\Rightarrow$ {definition of equality $SA.i = SB.j$}

$$h < N$$

End Remark .

*

Our remaining obligation is detailing "increase $h + i + j$ maintaining $P \land QA \land QB$" .

- Inspection of P shows that if $SA.i.h = SB.j.h$ —i.e. $A.(i+h)$ $= B.(j+h)$— an increase of h by 1 does the job. Note that QA and QB are maintained as well, because h does not occur in them.

- If $SA.i.h \neq SB.j.h$, we conclude $SA.i \neq SB.j$ or, more generally, $(\underline{A} p : 0 \leq p : SA.(i+p) \neq SB.(j+p))$.

More specifically we conclude for any p , $0 \leq p \leq h$:

$$SA.i.h > SB.j.h \land P$$

$\Rightarrow$ {definition of P , $0 \leq p \leq h$}

$$SA.i.h > SB.j.h \ \wedge \ (\underline{A}\,k : p \leq k < h : SA.i.k = SB.j.k)$$

$\Rightarrow$ {lexicographic ordering}

$$SA.(i + p) > SB.(j + p)$$

$\Rightarrow$ {definition of BB }

$$SA.(i + p) > BB \quad ,$$

from which we deduce, rewriting $SA.i.h$ and $SB.j.h$,

(4) $\quad QA \wedge P \wedge A.(i + h) > B.(j + h)$

$\Rightarrow$ {by the calculation above}

$$QA \ \wedge \ (\underline{A}\,p : 0 \leq p \leq h : SA.(i + p) > BB)$$

$=$ {renaming the dummy: $i + p = k$ }

$$QA \ \wedge \ (\underline{A}\,k : i \leq k \leq i + h : SA.k > BB)$$

$=$ {definition of QA }

$$QA(i := i + h + 1) \quad .$$

End $\bullet$.

And now we are ready to present "increase ..." :

$$\textbf{if } A.(i + h) \ = \ B.(j + h) \ \rightarrow \ h := h + 1$$
$$[\!]\ A.(i + h) \ > \ B.(j + h) \ \rightarrow \ i := i + h + 1 \ \{QA\text{, see (4)}\}$$
$$; \ h := 0 \ \{P\}$$
$$[\!]\ B.(j + h) \ > \ A.(i + h) \ \rightarrow \ j := j + h + 1; \ h := 0 \ \{\text{by (3)}$$
$$\text{and the preceding alternative}\}$$
$$\textbf{fi} \quad .$$

Note that in the last two alternatives, statement $h := 0$ is included so as to establish P (trivially), which may be falsified by assignments to i and j . Note, furthermore, that $h + i + j$ is increased by 1 , so that —see the guard $h < N \ \wedge \ i < N \ \wedge \ j < N$ of the repetition— the body is never executed more than $3 * N - 2$ times.

This completes the treatment of the algorithm.

The above presentation is a slightly adapted version of "Shiloach's Algorithm, taken as an exercise in presenting programs" by W.H.J. Feijen and A.J.M. van Gasteren, published in Nieuw Archief Voor Wiskunde. The major changes are the use of a somewhat more modern notation and the insertion of a missing step in one of the calculations.

Then and now, our incentives for writing the text were twofold. Firstly, though programs are designed and published that sometimes embody ingenious intellectual achievements, they are rarely presented in a fashion conducive to excitement. Secondly, this unfortunate state of the art in the field of presenting programs and justifying their correctness might very well be the reason why the mathematical community as a whole hardly recognizes computing science as a challenging branch of applied mathematics.

In order to narrow the gaps we wanted to demonstrate how a program can be presented clearly, concisely, and in all relevant detail by a suitable arrangement of the considerations leading to its development. As an example we chose an algorithm by Yossi Shiloach, published in Information Processing Letters.

We did so because on the one hand the algorithm was very ingenious yet presented in a way not revealing or doing justice to this ingenuity, and on the other hand we were convinced that the problem allowed a treatment that, firstly, would shed more light on how the algorithm might be designed and, secondly, would be clear and detailed enough to admit, for instance, presentation in an elementary course on programming.

In our presentation the two major design decisions become visible. The first one is the decision how to exploit P in the case $A.(i+h) \neq B.(j+h)$: the introduction of the concept of an ordering —where at first only equality and difference (of elements and sequences) were involved— , paves the way to exploit lexicographic order. The second design decision, which in all probability is a direct consequence of the first, is the introduction of AA and BB, and their usage to capture a whole bunch of inequalities $SA.(i+p) > SB.(j+p)$. Then, the contiguity of the bunch of inequalities $SA.(i+p) > BB$ suggests the introduction of invariant QA, with which all "inventions" of the design have been pointed out.

In our presentation we have aimed at a rather fine degree of detail. We did so for several reasons.

Firstly, it is the level of detail that most convincingly shows the correctness and efficiency of the algorithm, because it matches the fine-grained proof obligations like $P \wedge A.(i+h) = B.(j+h) \Rightarrow P(h := h+1)$ that the axiom of assignment brings about.

Secondly, the detail enabled us to be explicit about where exactly the design decisions come into the picture and, what is more, to justify why they have been taken. In other words, the fine degree of formal detail may provide heuristic guidance.

Finally, we have come to appreciate in general a homogeneous and rather fine degree of detail for mathematical arguments. That, as a result, the treatment above can be presented in an elementary course on programming, we consider an encouragement and an extra.

Part 1

Part 1

14 Clarity of exposition

We consider it beyond doubt that any mathematician wants his texts to be correct and clear. Yet they often are not: proofs contain quite a few "inaccuracies", many published algorithms are incorrect, and frequently a reader can only gain confidence in the correctness of an argument at the expense of a lot of time and intellectual effort.

We are particularly concerned with this problem in the area of program design: the demands of correctness and, hence, of ease of verification put on a program that has to operate impeccably a million times per day, say, without interference from its user, ought to be stiff, stiffer perhaps than for a mathematical theorem that is appealed to once a week, say, by a mathematician gifted with experience and common sense.

The desire to improve on the status quo has been our main incentive to investigate "clarity" as a topic in its own right. Compared to its recognized difficulty, however, clarity of written exposition receives remarkably little attention in the mathematical literature; Halmos's "How to write mathematics" is one of the rare examples of texts that do not exclusively have a didactics context but aim at providing guidelines to the practising mathematician as well.

Partly because of this scarcity of material and partly for the sake of comparison, we decided to investigate the topic by experimentation, with ourselves as guinea pigs: we wanted to learn to be more explicit about our standards of clarity and to learn how to meet and refine them.

Initially, we primarily experimented with (the exploitation of) the freedom of the writer that remains once the structure of the argument and the necessary ingredients have been chosen. The present chapter largely deals with findings of these experiments. Later experiments, in which the structure of the argument was affected by streamlining, are dealt with in the chapters on formalism and naming.

The main issues in this chapter are

- the division of labour between reader and writer;
- matters of arrangement, viz. of order and location of parts;
- disentanglement, i.e. the avoidance of local and global irrelevancies.

Before we proceed with the successive discussion of these three issues, one remark has to be made about the kind of mathematical texts to which our discussion is supposed to pertain. We certainly have in mind texts that are intended for publication or other forms of wider dissemination, but we also include the texts that a mathematician writes to record the results of his every-day work for, for instance, later usage, or the correctness proof that a programmer designs and records in the course of developing an algorithm. We shall rarely distinguish among such texts, because we consider the differences to a large extent marginal: the advantages of the coincidence of reader and writer quickly diminish with the passing of time.

*   *
*

14.0   The division of labour between reader and writer

The justification of the multitude of mathematical statements of fact that a proof usually contains, is a combined task of reader and writer, but it is the writer that, by his choice of what to include or exclude, dictates how much of that task is left to the reader.

On the one hand it is unavoidable that the reader takes his share, unless one resorts to first principles only; on the other hand statements of fact whose justification is wholly or largely left to the reader are a major source of error and of hard and time-consuming intellectual work from the reader's part.

Therefore, we decided to investigate how the writer could take a larger share of the work.

Remark . We were prepared to at least temporarily ignore objections often put forward against detailed proofs, such as that if a writer wants to be detailed, he will inevitably have to sacrifice brevity, or that if he is detailed, he is boring. As for brevity: as long as the length of a text is not a very good measure for the time needed to read that text, we are more concerned with techniques that may save the reader time than with ways to save paper and ink. And as for the danger of boredom: we would rather be boring than incorrect or unintelligible, and, in addition, no reader is forced to read everything a writer cares to write down; if an argument is presented in such a way that more detailed parts of the reasoning are easily recognized and skipped, readers with various opinions on what is boring can be served at the same time. In short, we consider the potential disadvantages mentioned above a separate and later concern: the art of omission has proved to be too difficult to be practised on the fly. In addition we doubt that such disadvantages inevitably occur. In fact they often don't.
End Remark .

*

The major problem for the reader that wants to fill in the details and gaps of an argument he is studying, is the size of the search space in which he has to find the ingredients for his proof.

Of course, his walk through the search space is not completely random. The (syntactic) structure of a demonstrandum already provides some hints: for instance, the structure of expressions $(\mu X : g(X).R)$ and $(\mu X : \mathcal{A}(g(X).T, R))$ suggests that for a proof of their (semantic) equality one may need the definition or derived properties of "μ", or "$\mathcal{A}$", or "g", perhaps even of all three of them.

As the above example shows, however, even a syntactic analysis alone already shows how much freedom of choice there is; and in addition other things may be needed, such as predicate calculus or concepts and properties that do not occur in the demonstrandum itself.

Therefore, we consider it a major task of the writer to keep the reader's search space small, i.e. to be explicit about what ingredients are to be used (or not to be used, for that matter). Such explicitness can be achieved in various ways. Consider, for instance, the following demonstrandum:

"Obviously, $(\forall_{u \in \mathcal{L}}(s - a)u \in C_b \;\Rightarrow\; u \in C_b)$
implies $(\forall_{u \in \mathcal{L}}(s - a)u \in C_0 \;\Rightarrow\; u \in C_b)$" ,

which at first sight has quite some degrees of freedom. If, however, the "Obviously" were replaced by "Since $C_0 \subseteq C_b$", one would have expressed very compactly that most of the internal structure is totally irrelevant. It helps a lot, and costs only one symbol.

Remark . We have observed that for some of the qualifications like "obviously", "trivially", "it is a simple matter to verify", and the like that we encounter, a more appropriate phrasing would be: "once the ingredients are known, it is a simple matter to verify".
End Remark .

Like the replacement of "obviously" in the example mentioned above, a simple hint like "predicate calculus" cuts down the search space quite effectively: with only two words it expresses that none of the properties of the atoms of a predicate are relevant.

To the extent that a hint's purpose is to narrow down the search space, it is hardly relevant whether such a hint is in some sense "trivial", or elementary, or familiar, or established earlier in the text: it is included first of all to express compactly what is irrelevant.

Therefore, the "trivial" may deserve explicit mention just as much as the non-trivial does. Similarly, explicit appeals to definitions are equally important as explicit appeals to theorems and lemmata. Yet the former occur much less frequently than the latter. In Apostol's "Mathematical Analysis", for instance, hundreds of definitions occur, each of them provided with a label, but explicit references to the definitions are extremely rare. Certainly there are situations in which the application of a particular definition is the only possible option, but then again there are many situations where that is not the case.

Remark . The addition of "trivial facts" and definitions can have yet another advantageous effect in the sense that it may simplify (the discussion of) heuristics. In, for instance, "On bichrome 6-graphs" [Chapter 2], the "trivial fact" that the total number of triangles is finite is the only thing necessary to realize the option of counting monochrome triangles by counting bichrome ones, a necessary step in finding the effective argument discussed in the chapter.

As another illustration, consider the proof quoted below, which has been taken from Courants "Differential and Integral Calculus". It demonstrates the existence of the limit of a (bounded, monotonic) sequence by proving the existence of an "accumulation point" and its uniqueness, and a, completely silent, appeal to some theorem relating unique accumulation points to limits. Definitions of limit and accumulation point are neither given in place nor explicitly referred to; the uniqueness proof constitutes the bulk of the argument.

> It is equally easy to see that *a bounded monotonic in-*
> *creasing or monotonic decreasing sequence of numbers*
> *must possess a limit.* For suppose that the sequence
> is monotonic increasing, and let ξ be a point of accu-

mulation of the sequence; such a point of accumulation
must certainly exist. Then ξ must be greater than any
number of the sequence. For if a number a_l of the
sequence were equal to or greater than ξ, every number a_n for which $n > l + 1$ would satisfy the inequality
$a_n > a_{l+1} > a_l \geq \xi$. Hence all numbers of the sequence,
except the first $(l + 1)$ at most, would lie outside the
interval of length $2(a_{l+1} - \xi)$ whose mid-point is at the
point ξ. This, however, contradicts the assumption that
ξ is a point of accumulation. Hence no numbers of the
sequence, and *a fortiori* no points of accumulation, lie
above ξ. So if another point of accumulation η exists
we must have $\eta < \xi$. But if we repeat the above argument with η in place of ξ we obtain $\xi < \eta$, which is
a contradiction. Hence only one point of accumulation
can exist, and the convergence is proved. An argument
exactly analogous to this of course applies to monotonic
decreasing sequences.

If, however, the definitions had been included, their similarities would
show why accumulation points enter the proof at all and suggest "each
accumulation point is a limit" as the remaining proof obligation, and
their differences would show how to fulfil that obligation. The resulting proof would be simpler and shorter, primarily because the proof of
uniqueness with its detail would disappear, as would the need to appeal
to the theorem. (The definition of "limit" says that for the limit of
a sequence each neighbourhood contains all but a finite prefix of the
sequence, and for an accumulation point each neighbourhood contains
infinitely many points (of the sequence), so the difference to be bridged
is that between "all but a finite prefix" and "infinitely many".)

End Remark .

*

In part, greater explicitness about the ingredients can be achieved in a simple and concise way, viz. if those ingredients themselves occur in the text. The relevant technique is well-known: it is labelling.

Yet the technique does not seem to be used very systematically. In the literature one may observe that, apart from definitions, theorems, and lemmata, formulae are about the only pieces of text that are labelled. (For example, we encountered the phrase "condition P implies" followed by a formula labelled "(0)" on a separate line. All subsequent appeals to (0), however, were in fact appeals to "P implies (0)", so labelling the latter would have been more accurate.)

What we would like to point out is that if labelling is restricted to formulae and one conducts a proof primarily verbally, a lot of necessary ingredients can only be referred to by repeating them; and since repetition is lengthy and laborious, the references tend to be omitted. (Sometimes even conditions appearing in the formulation of a theorem are nowhere mentioned explicitly in the proof.)

At the cost of only a few symbols, labelling facilitates greater explicitness and avoids the need for repetition. It can be used as refined as necessary: if, for instance, the conjuncts of a definition are appealed to separately, we give each of them a label, otherwise one label for the whole definition suffices. (For composite ingredients separate labelling is often the most handy choice.) In this way one can give very precise hints; in particular one can avoid stronger hints than necessary.

Finally, an additional advantage of a disciplined and sufficiently generous use of labels is that it reveals more of the structure of the argument; also it becomes more visible how frequently all kinds of statement of fact are used within one and the same proof, which may give an indication of how disentangled the proof is.

$$*\qquad*$$

Labelling helps being explicit in a concise way about proof ingredients that are contained in the text. A remaining question is to what extent an author is prepared to rely on theorems not proved —or even formulated— in his text and on concepts not explicitly defined, that is how self-contained he wants to make his arguments.

We have decided to attach great value to self-containedness. It is a personal choice, made in the first place for the convenience of the reader. (All too often we have found ourselves being put off by the multitude of undefined concepts in a mathematical text; we consider such thresholds unfortunate. Mathematical reasoning is in a way so universal that one would hope and expect that to a certain extent each mathematician should be able to read a well-written proof from an area that is not his own.)

The reader's convenience, however, is not the only goal that we have in mind. Since our interests are largely methodological, and our long-term aim is to learn how proofs can be designed in a systematic way, for us the important point about some concept or theorem —advanced or not— is not whether it does the job; the important question is how the shape of a demonstrandum can actually *lead* us to the concept or theorem that we need, even if it is completely new. In other words, we are looking for proof techniques that are more constructive than the appeal to knowledge is.

(We might add that there is still a long way to go. On the one hand we have, for instance, the concept of a "bichrome V" in the problem on bichrome 6-graphs [Chapter 2], which emerges as a result of carefully considered design decisions, but on the other hand there is the multiplication in the proof of existence of the Euler line [Chapter 3], which more or less comes out of the blue, and other examples.)

In view of these longer-term goals, we also strive for self-contained-ness on our own behalf: in doing so, we force ourselves to realize what exactly and how much "knowledge" we use. The beneficial side effect is that to nevertheless maintain brevity, we think twice before introducing another concept or a lemma that perhaps requires a long or complicated proof; as a result, the ultimate proof often turns out to be more detailed and yet simpler and shorter at the same time.

To become a little more concrete, for us proving something by declaring it a special case of some more general theorem, whose proof is omitted, is not as evidently acceptable as it would be, were establishing the validity of the demonstrandum our main objective. (In the treatment of the bichrome 6-graph-problem, for instance, the existence of a first monochrome triangle can be established by an appeal to the famous Ramsey theorem. Such a procedure, however, does not teach us much about the design of proofs; in addition, if in this case the formulation of Ramsey's theorem were included, as a matter of politeness to the reader, the resulting proof would even be longer than the one given in Exposition0 of Chapter 2.)

Similarly, that a concept is available in mathematics (with its properties) is not sufficient for us to use it in a proof, if it does not occur in the problem statement. As mentioned earlier, we would like the helpful theorem or concept to be suggested by the course of the proof development. (That such a helpful theorem or concept may very well turn out to be a generalization is a different matter; as long as not all premisses of a demonstrandum or components of a definition have been used, one is dealing with a generalization.)

We are trying to be frugal in our appeals to knowledge, but of course, we too do rely on "knowledge": we, for instance, freely use predicate calculus in our proofs, and we use the proof rules for programs in our derivations of programs. As for the predicate calculus, it is so little domain-specific and so widely applicable, that we consider it one of those calculi that one learns once and for all —like arithmetic—, to use it in one's daily work ever after. The proof rules for programs, though very few in number and still rather general, are of course more domain-specific.

We are the first to agree that striving for self-containedness, like the pursuit of other desiderata, is a matter of choosing a balance; one of our current interests is to see whether we can redress that balance. So we certainly do not mean to say that "knowledge" is of minor importance for doing mathematics, but for the time being we are playing a different game, a game we think worth playing.

We do not know how far we can get with it, but we are convinced that frugality in the number of concepts and simplicity of the theorems to be exploited are indispensable if we want to find more constructive techniques for the design of proofs. In the mean time, we find encouragement in the fact that striving for such frugality and simplicity has already often led to much simpler proofs.

In summary, the first reason for our striving for self-containedness is simplification of the reader's task, under the motto: "If it takes only a few lines and helps a lot, why not include a definition, a lemma, a proof, etc.". Then there is the more methodological reason, the interest in redressing balances, such as between domain-specific and less domain-specific knowledge used. Another balance that may need some redressing is that between the use of definitions and the use of derived properties of concepts. The discussion of Courant's proof some pages earlier, and the proof of existence of the Euler line in expository Chapter 3, for instance, illustrate how definitions, rather than being an extra burden on the length, can contribute to brevity and disentangledness of the proof.

*   *

So far we discussed being explicit about *what* is used in a proof. For clarity's sake we also want to be explicit about *how* it is used, and we want to do so in a somewhat more precise way than by, for instance, saying: "combining theorems A and B we may conclude". (Does "combine" mean taking the conjunction or not, are theorems A and B to be used with the same instantiation or not, is the order of application relevant?)

In other words, we are concerned here with the nature and the size of steps in reasoning that are left to the reader. Since filling in too large gaps in an argument is often an exacting and time-consuming activity for the reader, we aim at choosing steps fairly small.

Besides becoming simpler, arguments thereby also become more homogeneous in their grain of detail. That is desirable too. We, for instance, want to avoid that in one place the reader has to provide 10 lines of proof to fill a gap in the argument, while in another place a few lines suffice, one problem with such differences in size being that they are seldom apparent from the text itself. (An additional advantage of a fine, homogeneous degree of detail is that the length of a text becomes a more adequate measure of the time needed for reading it.)

A simple rule of thumb to keep steps in reasoning small is to write in such a way that the reader does not need pen and paper to verify the argument. Although not a very precise rule, it at least discourages us from leaving manipulations implicit that are too large or too difficult to do reliably and effortlessly by heart. We, in particular, have operations in mind that manipulate invisible intermediate formulae or results.

Compare, for example, the following proof with the alternative that follows it:

The invariance of Q, with Q defined as $p \mid x*m \wedge p \mid y*m$, under $x := x - y$ requires, besides the axiom of assignment, nothing but the distribution of multiplication over addition and the fact that, for all x and y, $p \mid (x - y) \wedge p \mid y \equiv p \mid x \wedge p \mid y$.

The alternative reads:

$$Q(x := x - y)$$
$$= \qquad \{\text{definition of } Q\}$$
$$p \mid (x - y) * m \wedge p \mid y * m$$
$$= \qquad \{\text{distribution of multiplication over addition}\}$$
$$p \mid (x * m - y * m) \wedge p \mid y * m$$

$$= \quad \{\text{additive property of division}\}$$
$$p \mid x * m \ \wedge \ p \mid y * m$$
$$= \quad \{\text{definition of } Q \}$$
$$Q \quad .$$

We think that the most important difference between the alternatives is the visibility, in the latter, of the intermediate expression "$p \mid (x - y) * m \ \wedge \ p \mid y * m$". Even in this extremely simple example, its presence considerably reduces the effort needed for verification, without making the proof longer.

Besides the not so very specific maxim of trying to make paper and pen superfluous for verification, we have a number of more fine-grained rules of thumb. They emerged in the course of our experiments, and we feel we have benefited greatly from trying to stick to them.

• Firstly, we use the rule of substitution of equals for equals as much as possible. The operation is attractive both from a logical point of view —because it is value preserving— and from the point of view of manipulation —because substitution is a simple operation, which can be performed "almost mechanically by the eye"— .

The occurrence of an application of this rule we always indicate by " $=$ " : schematically, for f a function we write

$$f.A$$
$$= \quad \{\text{hint why } A = B \text{, or } [A \equiv B] \text{ for predicates}\}$$
$$f.B \quad .$$

Note that an expression is a function of each of its subexpressions, so that the above boils down to replacing some, not necessarily all, occurrences of A by B. (In verbal arguments, " $=$ " is usually rendered as "equals", or as "means", "equivales", or "is equivalent to" in the boolean domain.)

The hint indicates the two expressions that act as the replaced expression and the replacing expression respectively. It may do so directly, as in

$$A \vee B$$
$$= \quad \{[B \equiv C]\}$$
$$A \vee C \quad ,$$

or by mentioning a symmetric rewrite rule that is to be instantiated, as in

$$q + (\underline{\text{MAX}}\, x : b.x : p.x)$$
$$= \quad \{\text{distribution of } + \text{ over } \underline{\text{MAX}}\}$$
$$(\underline{\text{MAX}}\, x : b.x : q + p.x) \quad ,$$

or by merely indicating "predicate calculus" or "calculus", if the rewrite rule is deemed sufficiently known and its applicability can be inferred sufficiently simply by parsing and pattern matching of the formulae.

Not only do we always indicate the use of substitution of equals for equals by " $=$ " , it is also the other way around: when we write " $=$ ", this almost invariably indicates a substitution of equals for equals. That means, for instance, that we rarely write

$$A$$
$$= \quad \{\text{hint}\}$$
$$B$$

—for $[A \equiv B]$— in a calculation, if that equivalence is to be established by separate proofs of $[A \Rightarrow B]$ and $[B \Rightarrow A]$. It also means that we allow ourselves to write

$$f.(A \vee B)$$
$$= \quad \{[A \Rightarrow B]\}$$
$$f.B \quad ,$$

although the relation in the hint does not have the shape of an equivalence: the presence of " $=$ " indicating substitution of equals for equals, the reader has no choice but to realize that by predicate calculus $[A \Rightarrow B]$ is equivalent with $[A \vee B \equiv B]$.

The possibility of this close connection between occurrences of " $=$ " and the use of substitution of equals for equals, has provided us with yet another reason for explicitly distinguishing steps that are value preserving from steps that are not.

• The most common steps that are not (necessarily) value preserving are implication ($\Rightarrow$) and follows-from ($\Leftarrow$), and similarly $\leq$ and $\geq$. Although formally they are redundant if equality and equivalence are available, they are included for the sake of convenience. (For a more extensive discussion, see the section explaining the proof format in Chapter 16.)

Just as equivalence and equality go together with equality preserving operations, i.e. with function application, in exactly the same way $\Rightarrow$ and $\leq$ go together with order preserving operations, viz. with the application of monotonic functions. Schematically, for monotonic f , we write

$$
\begin{array}{ll}
& f.X \\
\Leftarrow & \{\text{hint why } [X \Leftarrow Y]\} \\
& f.Y \quad .
\end{array}
$$

Again, the hint is to indicate the arguments to which the order preserving operation is applied.

The use of $\Rightarrow$ and $\Leftarrow$ works out so well because many of the logical operators are monotonic with respect to $\Rightarrow$, and, hence, their compositions are so as well: $\wedge$ and $\vee$ are monotonic in both arguments, $\Rightarrow$ is monotonic in its second argument, $\underline{A}$ is monotonic in its term, and $\underline{E}$ is monotonic in both term and range; and, in addition, negation is antimonotonic, and so are $\underline{A}$ and $\Rightarrow$ in their range and first argument respectively. Similarly, arithmetic operators like addition, maximum, and multiplication with a positive number are monotonic with respect to $\leq$.

•   As regards implication, there is a step that does not fit under the preceding heading. Consider, for instance,

$$x = y$$
$$\Rightarrow \qquad \{\text{Leibniz's Rule}\}$$
$$x \underline{\max} z = y \underline{\max} z \quad ,$$

which has no counterpart in arithmetic, i.e. a similar rule with " $\leq$ " instead of " $\Rightarrow$ " does not exist.   As a result, we consider such weakenings that are intrinsic to applications of Leibniz's Rule a separate type of step.

End • .

In the above we have enumerated three basic types of steps that we try to use whenever possible.

Our next rule of thumb is to be very cautious in combining such steps.   In general we avoid the combination of steps of different type, because then it is no longer possible to infer the type of a step from the symbol: a symbol " $\Rightarrow$ " or " $\leq$ " might stand for an order preserving step or for the combination of such a step with a value preserving step. The admittance of such combinations would require far more elaborate means to achieve the same degree of explicitness, the burden of which would have to be carried by the hints.

But even if we were to confine ourselves without exception to combinations of steps of the same type, which then could be inferred from the symbol used, we would be too liberal. The difference between combining two steps whose order is fixed and performing the steps one after the other is the absence in the former of the intermediate result, so that the second step of the combination requires the manipulation of an invisible expression. Some of such manipulations we try to avoid, viz. those that are nested substitutions, i.e. combinations where in the second step a term is replaced that entered the argument by the replacement performed in the first step.

     So much for our rules for combining steps. As mentioned before, they have greatly helped us, in deciding how to render our arguments, and in achieving the homogeneity and fine-grainedness of detail that we aim for. There is, however, no point in being dogmatic about following the rules. We sometimes *do* allow ourselves to combine nested substitutions, as in

$$A \lor B$$
$$= \quad \{ [B \equiv B \land C] \,;\, \text{distribution of } \lor \}$$
$$(A \lor B) \land (A \lor C)$$

or

$$Q \lor (\underline{A}\,x :: \neg Q \land P.x)$$
$$= \quad \{\text{distribution of } \lor \,;\; [Q \lor (\neg Q \land P.x) \equiv Q \lor P.x]\}$$
$$(\underline{A}\,x :: Q \land P.x) \quad .$$

(The semicolon in the hints indicates that the step consists of a succession of manipulations and it separates the hints for the individual manipulations.)

     Note, however, that combinations may require more specific hints, because the intermediate expression is not available for pattern matching. Compare, for instance, the latter step with the uncombined:

$$Q \lor (\underline{A}\,x :: \neg Q \land P.x)$$
$$= \quad \{\text{predicate calculus}\}$$
$$(\underline{A}\,x :: Q \lor (\neg Q \land P.x))$$
$$= \quad \{\text{predicate calculus}\}$$
$$(\underline{A}\,x :: Q \lor P.x) \quad ;$$

here we consider the hints "predicate calculus" together with the intermediate expression sufficient for deducing the rule applied, so that the gain of combining steps in this case is certainly debatable.

We terminate this section, dealing with explicitness about how proof ingredients are used, with two remarks. Firstly, a step like

$$A \vee (B \wedge C)$$
$$= \qquad \{\,\}$$
$$(A \vee C) \wedge (A \vee B)$$

not only appeals to distributivity of $\vee$ over $\wedge$, it also uses the latter's symmetry. Such an interchange of conjuncts might be considered totally ignorable in a context where a coarser grain of detail is admitted. In our context of finer detail, however, it constitutes a step. We have, therefore, adopted the discipline of avoiding transpositions like these unless and until they are truly needed.

Secondly, in the above we considered some types of step in reasoning without claiming completeness. We mainly concentrated on proofs in the calculational format, since we consider that format so convenient for achieving the explicitness we are after; among other things we do so because of the visibility of the " $=$ ", " $\Rightarrow$ ", and " $\Leftarrow$ ".

We also note how indispensable the hints, in their present format, are for our purpose. Of course, hints are not new; one, for instance, can see them used in proofs using natural deduction. Usually, however, the amount of space allotted to them is so small —like a third of a line or less— that it is hardly possible to be explicit about the nature of a step, let alone about combinations of steps or about the particular instantiation of a theorem used. We do not consider the use of long hints desirable, but such severe limitations of space are too inconvenient to be acceptable. Hence our generosity with the amount of space available for hints. Their incorporation as a fully fledged syntactic category so to speak —first suggested by W.H.J. Feijen— greatly simplifies such generosity.

*   *

*

14.1   On arrangement

Even if the contents of a mathematical text have been determined, the writer is left with quite a lot of freedom in how to arrange his material. How that freedom is best exploited may, of course, depend on what the text is about. If it is devoted to the development of some theory, concepts and theorems are studied in it in their own right; they are the subject matter; in the proof of some specific theorem, however, they are a means to an end: they are presented primarily because they are needed.   Such differences in rôle may well have an influence on, for instance, their positioning in the text.

We shall confine our attention to the situation of one well-specified goal, viz. to the task of proving a specific theorem. We do so because that is the situation we explored most; we have deliberately postponed larger-scale investigations, such as the development and presentation of conglomerates of theorems, because we think that the present task of exploring individual proofs needs to be dealt with first.

In our considerations we shall keep in mind that a text may be written for more than one purpose. On the one hand a writer may wish to communicate his result, as clearly and convincingly as possible, for the record so to speak. On the other hand he may also wish to convey information with a more methodological flavour, such as how the proof was designed, or why particular design decisions have been taken rather than others. We take the position that whenever it is possible to combine such multiple goals, that is well worth doing. (The reader of this monograph may notice that we, in fact, aimed at the combination in nearly all our little expository chapters.)

*   *

The one problem that we face when having to write down an argument is how to render an in general non-linear structure on a one-dimensional medium —our paper— : logically, we can view a proof as a directed graph, two logical components being connected if the validity of the one relies on the validity of the other, or —in the case of concepts— if the one is formulated in terms of the other. Because definitions —we ignore recursive ones— and arguments are supposed to be non-circular, the directed graph, having no cycles, represents a partial order.

The duty to linearize this partial order faces us with two types of considerations, viz. "ordered" ones and "unordered" ones. Firstly, we have to decide how to linearize an ordered pair. The choice is not irrelevant, because texts are read in one particular direction and not in the other; we shall assume in the following that reading is from left to right. Secondly, since the directed graph is usually far from complete, we also have to linearize unordered pairs.

We shall deal with unordered considerations first, and with ordered ones later. For the sake of simplicity we shall focus on (almost) tree-like arguments. After all they are not unusual in the case of one specific problem to be solved.

*   *

For the unordered concerns, consider a piece of argument of the form "$A \wedge B \Rightarrow C$", establishing C in three parts, viz. by a piece of reasoning that derives C from $A \wedge B$, a proof of A, and a proof of B. Let us also assume that the proofs of A and B are independent, i.e. that we are considering a tree with root C and leaves A and B.

With respect to C, A and B play exactly the same rôle; we can do justice to this symmetry by letting both or neither of their proofs precede C's proof. But there is no compelling reason to prefer order

AB to order BA.   All other things being equal, we have decided to choose an arrangement in which the distances between the formulations of A and B and their use in C's proof are minimal.   That means that the shorter of A's proof and B's proof is located in the middle. So if, for instance, B has the shorter proof, the only options left are ABC and CBA; the choice has to be made on the ground of "ordered" considerations.

We have formulated the above rule of thumb for statements of fact A, B, and C, but it is equally well applicable if A and B are concepts with their definitions that are needed in some statement of fact C, or in some definition C in the case of nested definitions.

If minimization of distance is our goal, that also means that if B is needed for some C but A is not, A will be outside the stretch of text that contains B and C.   In this way we can achieve that a lemma is as close as possible to its (first) use, and similarly for a definition and its use.   (It may be considered a disadvantage that in such an arrangement definitions are scattered all through the text rather than being concentrated at one place;   duplication in the form of a table of symbols or conventions or the like can make up for that.)

Remark .   The desire to minimize distances may bring forth strong hints for decomposition.   For instance, in "Not about open and closed sets" [Chapter 6] we introduce the concept "closed", by

$$T \text{ is closed } \equiv (\underline{A} x : aT.x : x \in T) \quad ,$$

separately from the definition of $aT.x$, because the two appeared to be used only in separate parts of the argument.   Here, like in many other cases, it is the introduction of nomenclature that makes such decompositions possible.
End Remark .

So much for our discussion of unordered concerns.

*

We now turn to ordered concerns, i.e. to the question of how to linearize a proof of B of the form "$A \Rightarrow B$", consisting of a proof of A and a derivation of B from A. Does the proof of A precede the use of A —in B's derivation— or is it the other way around? The predominant technique in the literature is not to use a result before it has been proved.

As a typical example, consider the proof by Courant quoted on page 94; although it is to establish the existence of a limit, limits do not enter the discussion until the very end of the argument in "and the convergence is proved"; similarly, the (comparatively long) proof of uniqueness of the accumulation point precedes the use, etc. .

A possible explanation for the predominance of the order proof-before-use is that maintaining this order guarantees the absence of circularity in the argument. But so does the reverse order —let us call it top-down arrangement— if used consistently. Truly tree-shaped arguments admit such a consistent use.

Top-down arrangement is by no means new. In classical geometry it is known, by the name "taking the demonstrandum as a starting point"; the familiar phrase "without loss of generality we may assume..." constitutes the first step of a top-down argument; and frequently the use of a reductio ad absurdum can be considered an effort to reconcile the wish to reason backwards with the obligation to use forward, implicational reasoning only. Except in these specific circumstances, however, top-down arrangement is quite rare. Its relative neglect in comparison to the reverse order, called bottom-up arrangement, was one of our reasons for paying special attention to it in our explorations. The two arrangements show a number of differences that we found certainly worth noting. The larger the argument concerned, the more pronounced the differences.

• Firstly, for the application of a result to be valid, only the *existence* of a proof of the result is relevant, the details of a particular proof are not (if they are, there is something wrong with the result as an interface). The top-down arrangement matches nicely with this situation: the proof with its overspecificity is postponed until the result has been used.

(One might object that it requires some faith from the reader to apply a result that has not yet been proved in the text, but faith is required anyhow: in the bottom-up arrangement the reader has to trust that the use of the result will follow.)

For simplicity's sake we have been considering only theorems in the above, but mutatis mutandis remarks similar to the above can be made about definitions as well. It is, for instance, very well possible to use (the name of) a concept in reasoning before a definition has been given. The earlier mentioned definition T is closed $\equiv (\underline{A}\, x : aT.x : x \in T)$ in Chapter 6 is used and manipulated before $aT.x$ has been defined, so that the reader is not burdened with details at a place in the text where they are not yet relevant.

In another argument we had to prove for some property P, that the number of elements of each finite non-empty set with property P is a power of 2. We did so by establishing that each set with property P and at least 2 elements has a subset of half its size and also satisfying property P. Since the validity of this pattern of proof does not depend on P, we felt free to postpone the presentation of all details of P until having presented the main pattern of the proof.

So much for a discussion of how top-down arrangement assists us in postponing detail until it is to be exploited. We note that, in the case of definitions, the introduction of nomenclature creates the helpful interfaces, in exactly the same way as the formulation of a theorem is the means for separating its use from its proof.

- Secondly, consider once more the tree-shaped argument "$A \wedge B \Rightarrow C$". In the bottom-up arrangement, the two independent proofs of A and B precede C's proof, so that the reader has to work his way through a number of unrelated pieces of reasoning until finally the simultaneous use of their outcomes —here: in the derivation of C— brings coherence into the text read so far. Particularly if the bottom-up arrangement occurs nestedly, for instance also in the proofs of A and B, the incoherence of such prefixes becomes more prominent. (A symptom is the frequent occurrence of turns of phrase like: "Next we prove" or "We also have" in an argument.) In contrast, top-down arrangement avoids such incoherent prefixes.

•   A third difference between the two types of arrangement is that top-down arrangement is more suitable to convey how the writer may have designed his argument. Although design seldom is a purely top-down affair, it tends to be more top-down than bottom-up. In the bottom-up arrangement, however, the reader starts reading "at the leaves of the tree", so to speak; the global structure — usually reflecting the main design decisions— remains largely invisible until almost the end of the proof.

Therefore, if it is considered important or instructive for the reader to pay attention to matters of design, top-down arrangement is more attractive, because the very first thing it confronts the reader with is the global structure of the argument.

For example, in top-down arrangement Courant's proof quoted on page 94 would have started with something like "The existence of the limit follows, by theorem ..., from the existence of a unique accumulation point; we prove the latter." . The reader would have known immediately that the main design decisions were to use accumulation points and to do so by appealing to a particular theorem. He could also have noticed at the earliest possible moment that Courant had not chosen the plausible alternative of proving that an accumulation point exists and that, in the context given, accumulation points satisfy the definition of "limit", the proof structure that we alluded to on page 95.

End • .

By these differences between bottom-up and top-down presentation, we have come to appreciate the latter, first of all as a service to the reader. He can see more clearly and earlier what the main design decisions are; detail is postponed, which makes skipping it easier. An added advantage is that top-down presentation more or less automatically invites the discipline of saying explicitly what you prove before proving it.

We have also come to appreciate top-down arrangement as a means to confront ourselves more explicitly with our design decisions, thereby prompting, at the earliest moment possible, questions about possible alternatives.

As soon as we had accepted top-down arrangement as a full-blown alternative to the more traditional bottom-up arrangement, and hence had to choose between the two all the time, we started encountering lots of arguments whose steps, when read in one direction, came as something of a surprise —a surprise only to be resolved further on in the proof— , while they were almost of the type "there is hardly anything else you can do" when read in the other direction.

We discuss one typical example of such an argument.

Example . We are to prove, for given predicates X and Y and function g on predicates,

$$[g.\textbf{true} \wedge Y \Rightarrow g.X] \quad ,$$

given

(0) $[f.(X,Y) \equiv Y]$;

(1) $[f.(X,g.X) \equiv g.X]$ for any X ;

(2) $[f.(X,Z) \Rightarrow Z] \Rightarrow [g.X \Rightarrow Z]$ for any X , Z ;

(3) f is conjunctive, i.e. for any P , Q , S , T ,
         $[f.(P,Q) \wedge f.(S,T) \equiv f.(P \wedge S, Q \wedge T)]$.

Note . Square brackets denote universal quantification over the domain on which the predicates are defined.
End Note .

Proof .

$$[g.\textbf{true} \wedge Y \;\Rightarrow\; g.X]$$

= {predicate calculus, preparing (2) with $X := \textbf{true}$ }

$$[g.\textbf{true} \;\Rightarrow\; \neg Y \vee g.X]$$

⇐ {(2) with $X := \textbf{true}$, $Z := \neg Y \vee g.X$ }

$$[f.(\textbf{true}, \neg Y \vee g.X) \;\Rightarrow\; \neg Y \vee g.X]$$

= {predicate calculus, preparing for f's conjunctivity}

$$[f.(\textbf{true}, \neg Y \vee g.X) \wedge Y \;\Rightarrow\; g.X]$$

= {(0) , the only thing given about Y }

$$[f.(\textbf{true}, \neg Y \vee g.X) \wedge f.(X,Y) \;\Rightarrow\; g.X]$$

= {(3) : f is conjunctive}

$$[f.(\textbf{true} \wedge X , (\neg Y \vee g.X) \wedge Y) \;\Rightarrow\; g.X]$$

= {predicate calculus, twice}

$$[f.(X, g.X \wedge Y) \;\Rightarrow\; g.X]$$

= { $[g.X \Rightarrow Y]$, by (2) with $Z := Y$, and (0) }

$$[f.(X, g.X) \;\Rightarrow\; g.X]$$

= {(1)}

true

End Proof .

We note that, if the above proof is read in reverse order, starting at "**true**", particularly the lower three steps look as rabbits pulled out of a hat: "**true**" can be rewritten in so many ways that the particular one chosen will always be a surprise; similarly, for the transformation of X and $g.X \wedge Y$ into conjunctions of two terms, there are many more possibilities than the particular one chosen in the third step from below.

In the order of presentation given here, however, each of these steps appears as a pure simplification.

End Example .

We note that the order chosen in the above example is only open to us by virtue of the availability of the symbol " $\Leftarrow$ " .

As the example testifies, in linear proofs of the type that establishes $P \equiv Q$ or $P \Rightarrow Q$ by a sequence of successive transformations transforming the one into the other, as in other calculational proofs, we are faced with the choice at which side to start. A rule of thumb that often helps is to start at the more complicated side, if one side can be recognized as such. Refinements of the rule of thumb are the following.

• A side with free variables not occurring at the other side counts as the more complicated one (as a consequence, constants **true** and **false** are the simplest of all). In other words, we rather recommend opportunity driven elimination than goal directed introduction.

• Often both sides contain as free variable an operator for which the rules of manipulation are strongly restricted; application of the operator to an argument for which the rules do not cater directly, counts as a complication, e.g. with an f about which little is known, expression $f.(\underline{A} X :: X)$ counts as more complicated than $(\underline{A} X :: f.X)$.

End • .

Having experienced that it is possible to avoid heuristic rabbits being pulled out of a hat, we made the personal decision to consider striving for the exorcizing of such rabbits a goal worth pursuing, not only in the presentation of proofs but also in the process of design. What started out as a mere matter of form —where to present what— has led to consequences of a methodological flavour.

With this discussion of top-down versus bottom-up presentation and their consequences we end this section on arrangement.

$$* \qquad *$$
$$*$$

14.2   On disentanglement

One of the qualities that we want our mathematical arguments to have is what we call "disentangledness"; by that we mean the avoidance of irrelevancies —be they global, i.e. avoidable altogether, or local, i.e. immaterial to part of the argument— . Our main reason for this desideratum is the wish to know and say precisely what is used and what is not.

If, for example, we have a complete finite graph whose edges can be partitioned into cycles of length 3, we want it to be explicit that the conclusion "3 divides the number of edges" has nothing to do with the cycles, only with partitioning into triples, and that the conclusion "at each node an even number of edges meet" has nothing to do with triples, but only with partitioning into cycles, and that so far the completeness of the graph has been totally irrelevant (it does, however, become relevant if one wants to use the above conclusions to derive properties of the number of nodes). Similarly (see "Not about open and closed sets"), if in an argument sets occur that are each other's complement, we want to be explicit about where this property is used and where it is not.

In a sense, the current section is both a companion to and a complement of the preceding sections. On the one hand, achieving clarity is a major goal to be served by disentanglement; on the other hand, while earlier the stress was on explicitness about what *is* used, here we stress the avoidance of what is *not* relevant.

Another reason for our conscious efforts to avoid irrelevancies is that, if successful, they may enhance the possibilities of generalizing the theorem and enable a more effective decomposition of the argument.

In this small section we shall discuss some, more macroscopic, means to achieve disentanglement, or, as it is also called, separation of concerns. Besides being separated, however, the concerns are also in some way connected by acting in the same proof or design. Therefore, a study of disentanglement is as much a study of how to choose one's interfaces.

We are not ready yet to present a collection of rules of thumb to achieve disentanglement; we shall largely confine ourselves to the identification of minor "slips" that may bring about entanglement.

$$*\qquad\qquad *$$
$$*$$

The perhaps most familiar instance of separation of concerns is formulating a theorem, so as to isolate the use of the theorem from its proof. The formulation itself forms the interface between the two parts and, hence, plays a double rôle. As a consequence of this double rôle, the formulation most convenient to use and the formulation most convenient to prove may not always coincide.

When faced with such a conflict, we have to choose which of the two rôles to favour. Because theorems tend to be used at least as often as they are proved, it is profitable to tune the formulation to smooth usage as much as possible, (and adjust it for the proof if necessary), i.e. to choose a formulation as "neutral" as possible.

The conflict manifests itself in a variety of ways. Without claiming completeness, we mention a few frequently occurring symptoms.

• The presence of nomenclature (in the formulation) that is useful in the proof but superfluous for the use of the theorem, as in "If a prime p divides a product ab then p divides either a or b", is such a symptom.

The phenomenon has been noted by others before, e.g. Halmos, yet can still be observed often.

•   For another symptom, the formulation of the type B hence C, consider the following formulation by Apostol, taken from "Mathematical Analysis A Modern Approach to Advanced Calculus". Apostol writes (Theorem 3-9) as follows:

> "(i) Each point of S belongs to a uniquely determined component interval of S.
>
> (ii) The component intervals of S form a countable collection of disjoint sets whose union is S."

Even without knowing what a component interval of S is, we see that (i) is a consequence of (ii). Combination of the two is, then, a clumsy interface. If (i) deserves the status of a theorem, for instance because it is needed in other proofs, it is best isolated as a separately formulated corollary; if it does not deserve that status, it need not be mentioned at all as a theorem. If, as in Apostol's case, it is included to indicate the structure of the proof, its proper place is in that proof.

•   Of two semantically equivalent formulations of a theorem, the one can be formally weaker than the other. For instance, that a relation R is symmetric can be expressed both as $(\underline{A}\,p, q :: pRq \equiv qRp)$ and as $(\underline{A}\,p, q :: pRq \Rightarrow qRp)$. For the *use* of R's symmetry, the stronger formulation is usually more convenient, because it allows substitutions of equals for equals; for *proving* the symmetry the weaker suffices, hence might in general be more convenient.

A similar avoidance of "formal redundancy" can be found in the familiar definition of a strongest solution (and mutatis mutandis of extreme solutions in general): "it is a solution and it implies all *other* solutions"; by predicate calculus, this is equivalent to "it is a solution and it implies all solutions" —the condition "other" has been omitted— . The latter characterization is more convenient to use, because the exploitation of its second conjunct does not require a case distinction: if, for

P the strongest solution, we encounter a demonstrandum $[P \Rightarrow X]$, we can reduce it to "X is a solution", independent of whether X equals P or not. (We note that in this particular example, the stronger formulation, rather than the weaker one, is more convenient for proving as well, because there is no simple way to exploit distinctness of predicates.)

•   Apart from the difficulty of choosing a formulation most convenient for usage and proof at the same time, the interface between proof and usage can also be inconvenient in other ways. Superfluous exceptions in the formulation of a theorem are an example: excluding, for instance, 1 in the theorem "Each integer greater than 1 has a unique prime factorization" is superfluous (because 1 is the unique product of 0 primes); even if the theorem is never used for the excepted instances, its application requires the superfluous checking that, indeed, the instantiation is within range. That we consider a (minor) inconvenience.

End • .

So much for some problems in the separation of the concerns of proving and using theorems. They all belong to the category *how* to choose an interface. A more difficult problem is *when* to choose an interface, in other words: to determine when a result that is used is worth being isolated. We shall not go into this topic here. We note, however, a number of familiar symptoms of missing interfaces, most notably the recurrence of similar pieces of reasoning or remarks to the effect that "the proof of B is similar to the proof of C" ; the most convincing and explicit way of expressing that two proofs are similar is by letting the corresponding demonstranda be instantiations of the same theorem. In a similar vein something is missing if we encounter a phrase like "the proof of B is a special case of the proof of C (replace x by y)" instead of "B is a special case of C" .

*   *

*

In the above we have mentioned some symptoms of lacking disentanglement. It is a topic of ongoing exploration. In individual arguments it is often quite clear what constitutes the disentanglement, or the lack of it, but as we said before, we are not ready yet to present a collection of rules of thumb.   One thing we do know, however, viz. that a fine grain of detail and top-down arrangement can contribute considerably to separate the irrelevant from the relevant.

15 On naming

It is impossible to do mathematics without introducing names; we need names to distinguish concepts and different instances of the same concept, or to indicate sameness by syntactic means; hardly a formula can be put to paper without the introduction of identifiers. There seem to be two main issues: what to name and how to name. An important distinction is whether a name occurs primarily in formulae to be manipulated or is to be used in a verbal context. We shall explore the two contexts separately, starting with the latter.

*   *
*

15.0 Names in a linguistic context

In a linguistic context, the first complication arises when a "meaningless" identifier is identical to a word in the language of the surrounding prose. Standard examples are the Dutch "U" —U is een samenhangende graaf— and the English "a" —a string beginning with a b— and "I" —because I may be empty— . On paper typographical means can be used to indicate which is which: the identifier can be made to stand out by the use of italics or extra surrounding space. In speech the distinctions are more difficult to render.

A greater complication arises when a normal word is given a

122

specific technical meaning.   Naming a set "checked" or "preferred", a boolean "found", a logical operator "defines", or a type "chosen" nearly precludes the use of these normal words in their traditional meaning. But even if an author explicitly states that some common words will only be used in this technical sense and consistently does so, the choice of such mnemonic identifiers requires great care, because their normal connotations can still invite confusion.

For instance, in a manuscript we encountered wires that could be "open" or "closed", but on closer scrutiny there appeared to be a third state, which was left anonymous.   Naming it "ajar" would have been faithful in the sense that transitions between "open" and "closed" only occurred via this third state, but no matter how faithful this name would have been, the nomenclature open/closed was confusing, because normally the two are considered each other's negation.   By a similarly confusing convention, in topology a set can be open and closed at the same time.   We also remember the confusion we once threw an audience into by introducing a state "passive" (for the machines in a network) that, instead of precluding all "activity", precluded only specific computations.

In situations like these, the use of colours is often convenient, e.g. "red" and "blue" for just exclusive states, "yellow" as the name of some property, and "white", "grey", and "black" to express a linear order.   Needless to say, the virtue of using colours immediately turns into a disadvantage if the colours denote non-exclusive states.

So much for the confusion that can be caused by the multiplicity of meaning of common words and by the author's inability to control which connotations they evoke.

*

A totally different concern is what we might call grammatical flexibility: does the term to be introduced have or admit the derivatives needed. As an example consider the adjectives "simple" and "complex".

Connected with the first we have the noun "simplicity", the verb "to simplify" and the noun "simplification"; for the other one we have the noun "complexity" and the verb "to complicate", but we lack the analogous noun "complification": the noun "complication" does not do, because it refers to the result and not the act of complicating.

Confronted with existing terminology, there may be no way out but to coin the missing term, e.g. "to truthify" in the meaning of "to effectuate the transition from false to true" in analogy with "to falsify". If one has to choose new terminology, the need for such coinages or the impossibility of forming a derivative had better be taken into account since they would make the terminology less convenient. (In this respect, the term "stack", for instance, is more convenient than the older "push-down list".)

*

Doing mathematics in a linguistic context involves a lot of verbal reasoning. This circumstance gives rise to other concerns.

Firstly, there is the following linguistic irregularity. In principle, prefixes and adjectives are constraining: a red dress is a dress and a wash-tub is a tub. Some prefixes, however, are generalizing and hence imply a hidden negation: a half-truth is not a truth, a semi-group is not a group, and a multiset and a fuzzy set are not sets. In natural language one can apparently cope with this irregularity, and even within mathematics words like "semi-group" and "multiset" are probably used as unparsed identifiers, but things become a little different with the weakly increasing sequence, which is not increasing. Having, then, the validity of "not (weakly increasing) $\Rightarrow$ not increasing", as opposed to the non-validity of "not (uniformly convergent) $\Rightarrow$ not convergent", complicates verbal reasoning.

The parentheses used in the above hint at another fuzziness pointed out by many before, viz. the natural use of "not", "and" and "or". The problem with "not" is that we cannot indicate its scope, vide

"not retired or disabled", and "not uniformly convergent".   The problem with "or" is that it is mostly exclusive, vide "It is a boy or a girl?",
but sometimes inclusive, vide "retired or disabled".   The trouble with
"and" is that it is sometimes conjunctive and sometimes enumerative,
vide "boys and girls".

Evidently, natural languages are not well adapted to cope with
these constructs. If new terminology intended for use in verbal reasoning is to be chosen, it pays to avoid them. In this respect it is worth
considering to give the complement of a named concept an equally positive name, not only because sentences with too many negations quickly
become confusing, but also because the negation, as in "unequal" versus
"equal" creates an asymmetry in the dichotomy that can be undesirable:
it suggests that the positively named term is the more fundamental one
and that the other one is a derived term with derived properties. Such
a suggestion of derivedness, however, may hamper reasoning.

That is why, for instance, the name "exclusive nor" —once suggested for equivalence— is unfortunate. In a similar way, the nomenclature marked/unmarked that we once introduced in a permutation
algorithm was unfortunate: elements began "unmarked" and ended up
"marked".   Besides destroying symmetry, this naming was also objec-
tionable because it nearly introduced marks as mathematical objects, in
the sense that it almost suggested state transitions to be performed by
adding or removing marks.   In terms of two colours, both operations
would simply be colour changes.

The simplest instance we know of concepts whose complements
deserve equally positive names is the trichonomy "less than", "equal to",
and "greater than". By introducing new names for their negations, viz.
"at least", "differs from", and "at most", one can avoid "not greater
than", and "less than or equal to", and "up to and including" (note the
enumerative "and" in the latter); similarly, "differs from" enables us to
avoid "not equal to" and "less or greater than".

The absence of complementary nomenclature for the above triple has an immediate analogue in the terminology for properties of sequences. Corresponding to "greater than", "less than", and "equal to", we have the "decreasing", "increasing", and "constant" sequence, respectively. Let us now consider what corresponds to "at least" and "at most". In the absence of this new terminology, "at least", for instance, is rendered by "not less than" or "greater than or equal to". For a sequence X with $X.i$ at least $X.(i+1)$ for all i, the corresponding characterizations a "not increasing" or a "decreasing or constant" sequence are both unacceptable, the former being too weak and the latter too strong. Mathematical parlance has programmed around it by strengthening the former to "monotonically not increasing" or calling it "weakly increasing" (a term discussed earlier). The whole problem is solved by introducing new adjectives, "ascending" and "descending", corresponding to "at most" and "at least" respectively. (Another alternative to be seen in the literature uses "increasing" and "strictly increasing", corresponding to "at most" and "less than" respectively, avoiding the anomalous "weakly", but introducing the term "increasing" in a different meaning.)

Yet another instance of the same lack of positive terms is the absence of such terms for "at least zero" and "at most zero", as opposed to the presence of the terms "positive" and "negative".

So much for naming in a linguistic context.

$$* \qquad *$$
$$*$$

15.1   Names in a formal context

First of all, in a formal context there should be no confusion as to which strings of symbols are names and which strings are more complicated expressions. Are P', P^*, and $\hat{P}$ identifiers, or are they the result of applying special functions denoted by $'$, $*$, and $\char`\^$ to P; is P'' a name or is it equal to $(P')'$? Is subscription a mechanism for name construction, e.g. $\triangle A_0 A_1 A_2$, or is it a special notation for function application, e.g. if after the introduction of infinite sequence $x_0, x_1, \ldots$ we encounter a reference to x_i. The difference may seem slight, but we note that in the case of A_0, A_1, and A_2 the name A is in the same scope available for other purposes, whereas in the case of $x_0, x_1, \ldots$ the name x is already in use as function identifier.

In this section we first explore the problem of how to name. We assume a relatively modest syntax for names introduced by the writer, say not much more than a sequence of letters and digits starting with a letter. (Even with such a restricted syntax, our freedom in choosing is usually large enough to avoid two-dimensional names with super- and subscripts or the use of various scripts; these more elaborate possibilities will be discussed as we come to them below.)

We take the position that formulae's main purpose is to be manipulated, and that, therefore, our choice of names should be guided by ease of manipulation. This position has a number of consequences, the most obvious being that a short name is better than a long one, not only for the writer but also for the reader, because a long name, by its size, complicates parsing. (Usually, we choose the largest symbols to denote operators with lowest binding power so as to assist the eye in parsing.) The so-called "self-explanatory" program variables whose use was made possible in higher-level languages were inconvenient for manipulation because of their length. They had, however, another disadvantage: it is usually much harder to find a sufficiently descriptive non-ambiguous name than to express properties of variables by predicates, the danger of committing "brevity by omission" being very high. In the case of a totally meaningless identifier it is obvious to both the writer and the reader that all relevant properties have to be stated explicitly.

A second consequence of the stress on manipulation is that it pays to choose one's dummies carefully. On the one hand one can try to choose them in such way that unnecessary substitutions are avoided, e.g. by naming an arbitrary solution of an equation by the same identifier as is used for the unknown. On the other hand, instead of exploiting equality of names, one may also have to exploit difference of names, e.g. if a theorem is to be instantiated one may avoid confusion by having the variables to be replaced named completely differently from what is to be substituted for them. (If presentation is concerned, the confusion might be avoided also by mentioning the instantiation explicitly, as by for instance $X, Y := X \wedge Y, X$; if design is concerned, however, things are different: when a formula to be manipulated is to be matched against potentially useful rules of manipulation, the pattern matching involved is simpler if the rules have completely different names.)

Thirdly, we often have to introduce names for groups of objects with some internal structure, e.g. pairs, cycles, a hierarchy, or reflecting a combinatorial state of affairs, e.g. vertices and edges of a triangle, and manipulation is simplified by choosing names that reflect such structure as much as possible. There are a number of well-known techniques, such as the use of primes, correspondence between small and capital letters, alphabetic nearness, alphabetic order, multi-character identifiers, and also the use of special alphabets.

Such techniques require a lot of care. Multi-character identifier AB , denoting the edge between vertices A and B , is a useful name, but the pair-wise relations in the set Xs , Ys , fx , and fy are more conveniently rendered by the set X, Y, x, y : formulae using them are shorter, and simpler for the eye; in the same vein we agree with Halmos (see "How to write mathematics") in preferring $ax + by$ to $ax_1 + bx_2$. Similarly, the names u, U, v, V in Chapter 4 reflect much more visibly the symmetry and pairings between (u, U) and (v, V) than, for instance, the choice $(x, y)/(u, v)$ would have done.

Reflection of symmetries , as in the above naming $u, U, v, V,$ is an important concern. In the bichrome–6–graph problem [Chapter 2] we could have named the six nodes P, Q, R, S, T, U , but such naming

totally obscures the symmetry among the nodes.   When, in the first
exposition, the six nodes are partitioned into two triples, we might have
used A_0, A_1, A_2, B_0, B_1, B_2 , but we would thereby have obscured
the symmetry among the A's and among the B's . That is why we
introduced the nomenclature AAA/BBB , which reflects the symmetry
within each triple by using indistinguishable names.

Special alphabets are frequently used to include type information
implicitly in the names chosen, e.g. Greek letters denoting reals next to
the Latin alphabet denoting rationals.   Apart from the typographical
problems that arise from the use of a variety of alphabets, such freedom
is often hardly necessary at all.   If one deals, as we do here, with one
mathematical argument, the number of instances of some type that are
needed simultaneously is usually small.   Rather than introducing a new
alphabet, one can therefore declare a few fixed identifiers to denote instances of a type, e.g. p, q, r for integers and B , C for sequences of
integers, or X, Y, Z predicates on some space and f, g, h predicate
transformers.   Even in ensembles of theorems this convention can sometimes be followed (see for instance the little fragment of ring calculus at
the end of Chapter 16).

So much for some of the problems of how to name.

*   *

Besides the question how to name there is the question what to
name.   As we hope to convey, the choice of what to name may influence
the structure of a proof;   thus naming is not just a matter of form: the
decision to give something a name (or not to do so) can be a genuine
design decision.

*

The first answer to the question what to name is: as little as
necessary.   The arbitrary identifier that is used only once is the simplest
instance of an unnecessary name, e.g. "Every integer N greater than 1

can be factored into a product of primes in only one way." (Courant and Robbins in "What is Mathematics?".) The remark has been made by others before, yet the phenomenon is a frequently occurring one. Superfluity can also be noticed in a statement like "In triangle ABC the bisectors of the angles A, B, and C respectively are concurrent" instead of "In a triangle the angular bisectors are concurrent". The statement displays a relatively mild symptom of superfluity: the text gets enumerative, possibly in combination with "respectively's", as can also be seen in, for instance, Exposition0 of Chapter 3, proving the existence of the Euler line, which says "[...] C is mapped onto the midpoint of AB, and cyclically A onto the midpoint of BC and B onto the midpoint of CA. Of course the images C', A', and B' respectively are such that $A'B'//AB$, etc. [...]". Figures in geometrical arguments are a rich source of superfluous names.

A more serious symptom of superfluity can be the occurrence of phrases like "without loss of generality we can choose", because it is then quite possible that an overspecific nomenclature has destroyed symmetries. Avoiding the introduction of the nomenclature may then involve more than just a cosmetic reformulation of the argument, and that is why we called the symptom more serious: it often indicates the possibility of designing a superior argument, such as replacing a combinatorial case analysis by a counting argument (as in the second treatment of the bichrome 6-graphs in Chapter 2) and other avoidance of case analyses.

*

Besides the superfluous name there is the missing name; being less visible, it can be more difficult to trace. Repetition, in the argument, of lengthy or similar subexpressions is the usual indication of this type of defective naming. Introducing names for the expressions serves brevity and hence convenience of manipulation.

But there is more to it than brevity. Having a name available enables us to show explicitly which part of an argument depends on the internal structure of the expression and which part does not, and thus offers possibilities for disentanglement and decomposition. For instance, in "Not about open and closed sets" [Chapter 6] we replaced Apostol's couple S and $U\backslash S$ by the couple S, T with S and T being defined to be each other's complement in some universe, which could then be left anonymous; this improves the separation of concerns, since any appeal to the relation between S and T now stands out more clearly. Postponing the detailing of $iS.x$ in "definition" $(\underline{A}\,x\,:\,x\in S\,:\,iS.x)$ of "S is open" has a similar effect: $iS.x$ acts as an interface between manipulations not depending on its internal structure and steps that do manipulate that internal structure. Yet another example is the proof of Helly's theorem [Chapter 9] in which the gradual introduction of more and more detail, in other words the postponement of detail, is effectuated by the introduction of nomenclature.

The treatment of Helly's theorem illustrates yet another concern for which the introduction of nomenclature can be essential, viz. the discussion of heuristics. If, for instance, in some context we have to prove that a predicate P is the disjunction of predicates from some subset of a universe V of predicates, the only thing we can do —if we don't want to pull a solution out of a hat— is to introduce a name for the subset or its characteristic predicate and restate the problem as solving an equation $[P \equiv (\underline{E}\,X\,:\,X\in V\wedge r.X\,:\,X)]$ in r (the characteristic predicate), so that manipulations of this demonstrandum can guide the design and make it explicit.

*

The third problem in naming is naming the wrong object; by necessity the boundary between this problem and the previous two —the superfluous name and the missing name— is somewhat vague because in a sense it combines the other two. We mention some common examples first.

• The first example is the superfluous argument of a function, i.e. naming a function instead of an expression (viz. the function application). A function's characteristic property is that it can be applied to different arguments, and if that flexibility is not needed and a function name, p say, always occurs in the combination "$p(s)$", it could be argued that a name for the combination, P say, would have been more appropriate. The phenomenon occurs often, and if we count subscription, etc. as notation for function application it occurs very often. It makes reading and writing unnecessarily laborious, vide

$$\text{``}\forall_{p(s)\in\mathbb{C}^m(s)}$$
$$B(s)p(s) \in \mathbb{C}^k_{\sigma,\infty}(s) \wedge \tilde{C}(s)p(s) \in \mathbb{C}^l_{\sigma,\infty}(s)$$
$$\Rightarrow E(s)p(s) \in \mathbb{C}^r_{\sigma,\infty}(s) \quad .$$

Let $p(s) \in \mathbb{C}^m(s)$ be such that premise is satisfied and let $\beta \in \mathbb{C}^+$. Then $B(s)p(s) \in \mathbb{C}^k_\beta(s)$ and $C(s)p(s) \in \mathbb{C}^l_\beta(s)$ ".

It may even take an unnecessarily large amount of space. On the page from which we took the above example (Journal of Mathematical Analysis and Applications 87, 1982, p.214), occurrences of the symbol string "(s)" constituted one fifth of all the symbols on the page.

Sometimes the choice of making the argument explicit is defended on the ground that in a few isolated instances the function is applied to other arguments. However, a notation for substitution could cater for these instances, e.g. $P^s_{s'}$, or $P(s'/s)$, or $(\lambda s.P)(s')$, or $P(s := s')$ for $p(s')$. (In the presentation of programs, for instance, the combination of named expressions, such as invariants, pre- and postconditions, and a notation for substitution is by now well-established; it is so ubiquitous there since the axiom of assignment, saying that "the weakest precondition such that $x := E$ establishes P is $P(x := E)$", is one of the core rules.) We sometimes employ notation $P^s_{s'}$, but mostly write $P(s := s')$. In view of the above connection with assignment statements one also sees $(s := s'); P$.

Sometimes the need to express universal quantification (over s) is felt as a necessary reason for introducing $p(s)$ rather than P :

($\underline{A} s :: p(s)$) looks better than ($\underline{A} s :: P$), because one wonders whether the dummy is arbitrary or not. In a context, however, where quantification over s is the order of the day, great economy of expression can be achieved by denoting universal quantification over s by a reserved bracket pair, e.g. $[P]$ instead of ($\underline{A} s :: p(s)$).

• A second example of the difference that naming the right concept may make is what we might call choosing the right coordinate system. Analytical geometry shows a lot of instances; a very simple one, for instance, is the following: in the derivation of an algorithm computing the coordinates (X, Y) of a pixel on the screen, subexpressions $(X - X_0)$ and $(Y - Y_0)$ occurred all over the place; naming these expressions, that is calling the pixel coordinates $(X0+x, Y0+y)$ would have circumvented this. Another example in a similar vein is the named predicate expression that only occurs with a negation sign; naming the negation of the expression would have been more appropriate.

• Naming a subset versus naming the characteristic predicate, is a next example. Given a predicate p, on natural numbers, say, we can define set P by

$$P = \{n : p.n : n\} \quad ;$$

conversely, given the set P we can define predicate p by

$$(0) \qquad p.n \equiv n \in P \quad .$$

In view of this one-to-one correspondence, we need only one of the two. We have observed that the majority of mathematical texts use P. In view of the similarity of the left- and right-hand side of (0) the choice seems almost irrelevant. But it isn't, because the logical connectives —at least as we use them— have been developed better than the set-theoretical operators. The latter suffer from the constraint that each element of the result is an element of at least one of the operands —$\cup$, $\cap$, $\setminus$ (or $-$), and symmetric difference $\div$— or from the anomaly that the result is not a set (but a boolean) —$\subseteq$ and $=-$.

In particular the first constraint makes set calculus less convenient: the hesitation to accept the presence of some universe precludes the use of the complement operator and the exploitation of the set-theoretical equivalent of equivalence as a connective. An illustration of the possible inconveniences can be found in the Aside in "Not about open and closed sets" [Chapter 6]. A further illustration arises from an expression like $[a \equiv b \equiv c \equiv d]$ which is completely symmetric in a, b, c, and d; its set-equivalent, for instance $A \div B = C \div D$ or $A \div C = B \div D$, inevitably has to break that symmetry. Also, it is not very attractive that

$$A \cap (B \backslash C) = (A \cap B) \backslash C$$

is valid, whereas

$$(A \backslash B) \cap C = A \backslash (B \cap C)$$

is not. In general, we have found the application of the logical connectives usually much more convenient.

• The final subject of naming we address is the use of subscripts. We do not know how Cayley and Sylvester invented the matrix calculus, but we can imagine that at some time they got tired of writing

$$
\begin{aligned}
A_{11} \cdot x_1 + A_{12} \cdot x_2 + \cdots + A_{1n} \cdot x_n &= y_1 \\
A_{21} \cdot x_1 + A_{22} \cdot x_2 + \cdots + A_{2n} \cdot x_n &= y_2 \\
\vdots \qquad\qquad \vdots \qquad\qquad\qquad \vdots \\
A_{n1} \cdot x_1 + A_{n2} \cdot x_2 + \cdots + A_{nn} \cdot x_n &= y_n
\end{aligned}
$$

over and over again, and one of them suggesting: "Why don't we write $A \cdot x = y$?", thereby opening the way for the economic expression of rules for manipulation and calculation. In a much more modest context we did the same when towards the end of Chapter 16 we showed a fragment of a calculus for permutations: rather than "naming" each individual element of a cyclic permutation — such as in $(a_1 a_2 \ldots a_n)$—, we avoided the subscription by naming the sequence, with some change of notation writing $[A]$. Only then did the expressions become amenable to manipulation.

In the expositional chapters, the avoidance of subscription and the advantages of doing so form a recurring theme. Besides other goals such as avoiding overspecificity and maintaining symmetries, the main reason why we try to be as frugal as possible with the introduction of subscripted variables is manipulative convenience. In many cases they have proved to be quite avoidable.

End • .

With the above remarks on subscription we end this exploration of how to name in a formal context. We have stressed that in such a context manipulative convenience is the driving force behind the choices to be made. In this respect, inadequate naming tends to manifest itself fairly clearly. We are the first to admit that in general it is less clear how to remedy the situation: as some of the examples show, the concept to be named might need to be invented. We do hold the belief, however, that in the process of such invention considerations of manipulative convenience can be of considerable heuristic value.

16 On the use of formalism

> "[Symbolisms] have invariably been introduced to make things
> easy. [...] by the aid of symbolism, we can make transitions
> in reasoning almost mechanically by the eye, which otherwise
> would call into play the higher faculties of the brain. [...]
> Civilisation advances by extending the number of important
> operations which can be performed without thinking about
> them."
>
> Alfred North Whitehead
> "An Introduction to Mathematics"

Formulae can be manipulated and interpreted. Whitehead apparently
advocates manipulation without interpretation. Many mathematicians
do not believe in it, nor do they believe in a combination of the two.
Halmos represents this opinion when, in "How to write mathematics", he
writes: "The symbolism of formal logic is indispensable in the discussion
of the logic of mathematics, but used as a means of transmitting ideas
from one mortal to another it becomes a cumbersome code. The author
had to code his thoughts in it (I deny that anybody thinks in terms of $\exists$,
$\forall$, $\wedge$, and the like), and the reader has to decode what the author wrote;
both steps are a waste of time and an obstruction to understanding." .
Somewhat further on he says about proofs that consist in "a chain of
expressions separated by equal signs", that these, too, are encodings, and
suggests that authors "replace the unhelpful code that merely reports the
results [...] and leaves the reader to guess how they were obtained." .

As for the chain of expressions separated by equal signs, the question is whether it is the presence of the chain or merely the absence of hints that is to be blamed for the "unhelpful"ness of the "code"; we think it is the latter. As for constant decoding, we agree with Halmos that it, indeed, is a nuisance to be avoided as much as possible. Unlike Halmos, however, we try to do so by following Whitehead's ideal of avoiding decoding rather than by avoiding formalism.

Although in the light of history, the scepticism as voiced by Halmos is not unjustified —Whitehead's own work may be related to it— , as computing scientists we had, at the outset of our investigations, a compelling reason to hope that Whitehead's ideal would be feasible, and some encouraging experiences to support that hope.

The compelling reason was that programs are formulae —and long ones, for that matter— whose interpretations have proved to be very hard to work with: interpretation of a program as the set of computational histories that can be evoked by it, forces one to reason about a combinatorially complex set, which experience has shown to be hard to do without error.

The initial encouraging experiences included, for instance, the apparent feasibility of formally deriving now canonical algorithms like the Binary Search or Euclid's algorithm for the greatest common divisor from their specifications. Also, activities like making compilers had familiarized the computing scientists with the manipulation of uninterpreted formulae.

Thus, besides conciseness and precision, manipulability became an important property of formalisms to us; investigation of the requirements that formulae and their manipulations may have to meet in order to make interpretation unnecessary then became an explicit concern.

There are two major reasons why manipulation, and in particular manipulation without interpretation, is so important to us: the first is that we want reasoning about programs and (a-posteriori) verification of proofs to be as simple as possible, and the second is that we are exploring how the structure of formulae in combination with a body of rules of manipulation can guide the design of proofs and help us to be explicit about design decisions as much as possible.

We have two main concerns in this chapter: first we deal with what is necessary to make manipulation without interpretation feasible at all, next we discuss what can be done to make it convenient and non-laborious as well.

*   *

*

16.0   Manipulation without interpretation

Since one of our goals is to use and manipulate formulae without having to interpret them, we discuss necessary conditions for achieving this goal first.

A preliminary remark to start with. If we do not wish to rely on interpretation in terms of an "underlying model", we have to rely on the symbols written down, because they are now the only thing we have got. Consequently, every symbol counts, and accuracy becomes so vital that it has to become second nature. We make this remark because failure to accept this consequence has led to disappointing experiences with formula manipulation. In what follows we take the acceptance of the consequence for granted.

This having been said, the first thing to stress is that, if the formulae have to carry the load, they and nothing else have to determine which are the legitimate manipulations. This means that the formulae have to be unambiguous, that is do not admit semantically different parsings. We shall discuss ambiguity in some more detail.

Non-ambiguity of the formulae

In order to illustrate how quickly one is in danger of introducing ambiguous formulae, we consider an example from "Ramsey theory", by Graham, Rothschild, and Spencer.

Example

Graham et al. introduce the following two definitions (in a somewhat different phrasing). (We invite the reader to compare their definitions without bothering about how concepts occurring in them are precisely defined.)

Definition0 . We write $n \to (l)$ if
given any colouring of $[n]^2$ with 2 colours, there is a set T,
$T \subseteq [n]$, $\#T = l$, such that $[T]^2$ is monochromatic.
End Definition0 .

This definition is almost immediately followed by

Definition1 . We write $n \to (l_1, ..., l_r)$ if
given any colouring of $[n]^2$ with r colours —1 through r— there is
an $i : 1 \le i \le r$ and there is a set T,
$T \subseteq [n]$, $\#T = l_i$, such that $[T]^2$ is monochromatic of colour i.
End Definition1 .

Graham et al. remark that now $n \to (l)$ and $n \to (l, l)$ denote the same thing. They fail to note, however, that now $n \to (l)$ is ambiguous: interpreted according to Definition0, it refers to colourings of $[n]^2$ with 2 colours, interpreted according to Definition1 with $r = 1$, it refers to colourings of $[n]^2$ with 1 colour. Because the value of $n \to (l)$ may depend on this choice, the authors have introduced an ambiguity. Apparently, we have to be very careful whenever we wish to introduce new notation.

End Example .

Standard mathematical notation itself contains quite a number of ambiguities as well. We discuss some common ones.

•   Consider sequence "expressions" $(1, \ldots, k)$ and $(x_1, \ldots, x_k)$. They look similar, but they are not. If we substitute $x_1 := 1$ and $x_k := 7$, do we get the sequence $(1, \ldots, 7)$, i.e. $(1, 2, 3, 4, 5, 6, 7)$, or something else —like $(1, 7)$ if $k = 2$— ? Usually, the answer will be: "something else", but then the conclusion is that $(x_1, \ldots, x_k)$ is a formula for which substitution is not allowed, a rather unusual type of formula.

•   For a function f, expression $f^2 x$ is used to denote two different things: in "$\sin^2 x + \cos^2 x = 1$" we know it to stand for $(f(x))^2$, with sin and cos for f, but it is also used to denote repeated function application, i.e. to stand for $(f \circ f)(x)$ —or, $f(f(x))$— . Similarly, $f^{-1}(x)$ sometimes stands for $(f(x))^{-1}$ and sometimes denotes application of f's inverse to x. Here, the desire to save a few parentheses has led to ambiguity. In both cases, the first convention is the anomaly. We note that part of the problem is caused by the exponent notation, because it leaves the associated binary (associative) operator implicit: from expression a^n we cannot infer which operator is applied repeatedly on a.

•   Even in programming languages, ambiguities have been introduced. In an early version of ALGOL 60, an expression of the form **if** B **then** S could be two different things: an incomplete prefix of a statement **if** B **then** S **else** T, or an abbreviation of the complete statement **if** B **then** S **else**, which terminates with the invisible statement denoting skip. As a consequence, **if** $B0$ **then if** B **then** S **else** $S0$ was syntactically ambiguous. (Although a pair of parentheses would have resolved the ambiguity, it was not obligatory.)

•   Another source of ambiguity in formulae is the way in which dummies are treated. There are two important concerns here. Firstly, it should be visible which "variables" are dummies, and, secondly, their scopes should be clear.

As regards notations that obscure the identity of the dummy: there are quite a few around. The usual set notation is, in principle, explicit about the identity of the dummy and hence OK: in $\{i \mid B.i\}$, the "i" in front is the dummy. The problem is that it also denotes the element: to denote, with this convention, the set obtained from the previous one by squaring each element, we have to write something like $\{j \mid (\underline{E}i : B.i : i^2 = j)\}$. This is so clumsy that occasionally the set is denoted by $\{i^2 \mid B.i\}$. Thus we may write $\{i^2 \mid i > 2\}$ and $\{i^3 \mid i > 3\}$, but by the time that we find ourselves considering $\{i^j \mid i > j\}$, we are lost: we have obscured the identity of the dummy.

Had we written $\{i : B.i : i^2\}$ to start with —employing a notation that explicitly mentions the dummy in the position preceding the first colon— , we would have ended up with $\{i : i > j : i^j\}$, which definitely differs from $\{i, j : i > j : i^j\}$. (These ambiguities do occur. Lately we saw a text that contained quite a few instances of "$\underset{i \neq j}{\wedge}$" (with "$\wedge$" denoting universal quantification); in half the number of cases i was the only dummy, in the remaining cases j was a dummy as well.)

Explicit declaration of the dummies also avoids ambiguities in other quantified expressions, such as summations or "maximizations". It, for instance, avoids the (usually tacitly assumed) distinction between $\sum_{i=j}^{n} \ldots$ and $\sum_{j=i}^{n} \ldots$ (the assumption being that the left side of the symmetric(?) equality denotes the dummy), or the ambiguity of $\underset{a<b<c}{\max} (a^2 - b)$. Similar problems are avoided if the unknowns of equations are declared explicitly, so that, for instance, the distinction between $(x : x^2 + y^2 = 10)$ and $(x, y : x^2 + y^2 = 10)$ is made visible by syntactic means.

So much for avoiding obscurity of the identity of the dummy; let us next consider matters of scope. One sometimes finds universal quantification denoted by prefixing with "$(\forall x)$". Without explicitly stated priority rules, it is not clear how $(\forall x)P.x \vee Q$ and $(\forall x)P.x \wedge Q$

are to be parsed. For the first formula that is in a way OK, because its two parsings

$$((\forall x)P.x) \vee Q \quad \text{and} \quad (\forall x)(P.x \vee Q)$$

are semantically equivalent; the parsings of the second formula, however, viz.

$$((\forall x)P.x) \wedge Q \quad \text{and} \quad (\forall x)(P.x \wedge Q) \quad ,$$

are only equivalent if the range of x is non-empty —for an empty range they yield Q and **true** respectively— . Such problems can be avoided by the use of an obligatory pair of parentheses that delineates the scope of the dummy.

In summary: for all sorts of quantified expressions we employ a notation that explicitly declares the dummies, expresses their range as a boolean expression, and, thirdly, contains a term expressed in terms of the dummies, with a pair of parentheses delineating the scope of the dummies.   Thus we achieve non-ambiguity and at the same time a greater homogeneity than the established notations possess.

End • .

So much for some of the sources of ambiguity in formulae, and the ways in which we avoid them.

End Non-ambiguity of the formulae .

So far we have only discussed how the shape of *formulae* could be chosen so as to help avoid the need for interpretation. The other necessary condition is that the shape of the rules of manipulation be appropriate as well. By that we mean that we would like the rules of manipulation to be cast in such a form that applying them can be as mechanical an operation as substitution.

To experience the differences, consider the following rule. Expression $(m ; p, s : q)$ is defined to denote "an array equal to m except that $m[p].s = q$ ". In the paper containing this rule it was required to reformulate expression

$$(m;\, u, s : v)\,[(m;\, u, s : v)[w].s].s \;=\; v$$

in terms of m, by using the definition. The resulting formula turns out to be

$$(u = w \,\wedge\, (u = v \,\vee\, m[v].s = v))$$

$$\vee$$

$$(u \neq w \,\wedge\, (u = m[w].s \,\vee\, m[m[w].s].s = v))\quad.$$

With the definition of $(m ;\, p, s : q)$ given above, such a reformulation can hardly be a mechanical operation. With a definition like, for instance,

$$(m;\, u, s : v)\,[w].s \;=\; \textbf{if } u = w \rightarrow v \;\;[\!]\;\; u \neq w \rightarrow m[w].s \textbf{ fi}\quad,$$

the rewrite can be performed without thinking, by applying rules of predicate calculus and of the calculus of conditional expressions **if ... fi** . (The resulting formula is fairly complicated; therefore, it seems all the more necessary to keep the manipulations explicit and simple.) By its shape, the second definition is more geared to manipulation in a calculation.

So much for the shape in which to cast the rules of manipulation. The next question is which rules to include explicitly. For the sake of expedience, we may omit what can safely be assumed to belong to the "common knowledge" shared by reader and writer, e.g. the rules of integer arithmetic. In less familiar domains, however, it usually pays to be explicit. If one wants to play the game syntactically, one has to provide enough rules to do so. One of the added advantages of such syntactic reasoning is that whenever one is forced to appeal to the interpretation, the model, the conclusion is that there is some theorem that needs to be formulated.

*

In the above we have discussed a number of properties that formulae and their body of manipulations need to have, if manipulation without interpretation is to be feasible at all. The outcome of our experiments is that if a formalism satisfies such properties, it is, indeed, possible to render and verify proofs in that formalism without the need for interpretation (or "decoding", in Halmos's terminology). Whether besides being possible it is also convenient is a different matter, to be addressed in the next sections.

There is one important point about non-interpretation, however, that we have not discussed yet; that is Halmos's suggestion that interpretation is needed for heuristic purposes ("I deny that anybody thinks in terms of ∃, ∀, ∧, and the like"). We disagree. Far from declaring it the only source of inspiration, we think that using the syntactic structure of formulae as the main source of inspiration in the design of a proof or a program is an underexploited technique. Developing agility in the use of this technique was (and still is) one of the goals of many of our experiments. The outcomes are encouraging. (The formal derivation of a proof for the invariance theorem in Chapter 10 is an illustration.)

*   *

*

16.1   On convenience of manipulation

In the preceding we discussed necessary conditions for making calculational reasoning, syntactic manipulation, feasible at all. In the following we shall be concerned with the question of how manipulation, in addition to being possible, can be made convenient as well.

The notion of "convenience" is, of course, closely related to the mechanisms that one has available for manipulation. Mechanized formula manipulation as performed by a computer is quite different from manipulation with pen and paper, and the latter in its turn even distinctly differs from using a blackboard with a piece of chalk in one hand and an eraser in the other.

We have chosen pen and paper for doing our manipulations with, and therefore, repetition and lengthiness are among the evident sources of inconvenience. We shall discuss the avoidance of such laboriousness later. First we want to address what we consider the most important way to achieve convenience of manipulation, namely keeping the rules simple.

*   *

Parsing

The two operations that we perform most frequently when manipulating formulae are substitution of an expression for a variable and replacement of an expression by another one, of the same syntactic category. That means that, while manipulating, we are parsing our formulae all the time. It is, therefore, a matter of considerable convenience if the parsing algorithm is as simple as possible.

This desideratum is, first of all, a second reason for avoiding ambiguity (besides the earlier mentioned one, viz. avoidance of the need for interpretation of the formulae).

Secondly, we note that the introduction of infix operators is a technique that complicates parsing, the problem being that infix operators invite the introduction of priority rules so as to reduce the number of parentheses needed. Such priority rules complicate parsing, and hence, substitution: substitute, in the product $x*y$ the expression $2+3$ for x. Slight though this complication may be, it never hurts to remember that it is there. In addition, when one introduces an infix operator, it may be difficult to foresee which priority will reduce the number of parentheses most effectively. (We also know from experience and observation, that beginning by leaving the priority open has the danger of ending up with ambiguous formulae.)

For associative operators, however, infix notation is attractive. There are many of these, such as addition, multiplication, greatest common divisor, maximum, minimum, conjunction, disjunction, equivalence, concatenation, relational or functional composition. It is much nicer to write

$$a \underline{\max} b \underline{\max} c$$

than

$$\underline{\max}(a, \underline{\max}(b, c)) \quad \text{or} \quad \underline{\max}(\underline{\max}(a, b), c) \quad .$$

Remark . By omitting parentheses in the first alternative, we introduce a syntactic ambiguity without, however, introducing a semantic ambiguity; it does not matter where the parentheses go; therefore, they are omitted for convenience of manipulation.
End Remark .

The second notation not only has the disadvantage of having more parentheses, it also forces one to make a totally irrelevant choice. An inevitable consequence is that calculations may be longer than necessary, because one has to include steps that pass from the one choice to the other. (If a functional notation is preferred, the better alternative is to write $\underline{\max}(a, b, c)$ —i.e. a function applied to a non-empty list— .)

So much for infix operators. The third complication of parsing that we discuss here is the admission of context-dependent parsing, i.e. parsing that depends on the type of the operands. The problem with context-dependent parsing is, first of all, that it complicates the demonstration that the formalism is unambiguous —a demonstration that is required at each, even temporary, extension of the grammar— . Secondly, it complicates the use of the formalism.

For boolean equivalence, for example, we could use the normal equality sign. We could, in addition, decide to give relational operators on the integers greater binding power than the boolean operators —as anyone would be tempted to do when working with logical formulae in which such operators occur in boolean primaries— ; and we are also tempted to give boolean equivalence about the lowest binding power among the logical operators. Then, however, a formula like

$$a = b \wedge c$$

cannot be read without knowledge of the types of the operands.

The above was just an example. The introduction of invisible operators is the most common way of introducing context-dependent parsing —if not down-right ambiguity— . (Compare $3x$, $3\frac{1}{2}$, and 32 . We are so familiar with these notations that we hardly notice the anomaly any more, although occasionally we may encounter, say, "11/2" without having a clue whether it stands for 1.5 or for 5.5. Yet, our own familiarity with these anomalies does not seem enough reason to burden new generations of school children with them.)

The operators most often left invisible are multiplication, concatenation, and function application. The resulting overloading of juxtaposition has more unpleasant consequences.

It virtually precludes the use of multi-character identifiers, and the latter are, indeed, rather rare in mathematical texts. To compensate, subscripts and characters from outside the Latin alphabet are very common. (It is, for instance, not unusual to see the vertices of a triangle denoted by A_1 , A_2 , and A_3 , where $A1$, $A2$, and $A3$ would have done just as well.) Similarly, besides the Latin alphabet we can see Greek, Gothic, Hebrew, and large script capital letters in use, while the admission of even two-character identifiers only would already give us 52 new "alphabets".

The wide typographical variety of conventions like subscription and the use of many character sets has always made typing and printing mathematical texts a difficult, costly, and error-prone activity. In recent years, it has created the need for sophisticated text editing systems, that are apparently difficult to design and difficult to use.

To avoid the consequences of the use of invisible operators, one has to introduce symbols for them. For function application this monograph uses the low infix dot suggested by Dijkstra, which is given the highest possible binding power and, hence, a size as small as possible. So far, experiments with it have been favourable. (It has the added advantage of saving parentheses around arguments consisting of a simple variable.) For concatenation of sequences, various notations are in use, among others ++ (see Richard Bird and Philip Wadler, "Introduction to Functional Programming").

End Parsing .

*   *

There is more to be said about keeping the rules simple. The choice of how many rules to include explicitly in a formalism is a matter of delicate balancing: if there are too few, proofs will be long, if there are too many, the multitude of rules will burden the memory and complicate design. In view of the latter complication in particular, techniques to bridle the multitude are apparently most welcome. Maintaining homogeneity is one of these, maintaining symmetry another.

Homogeneity

Doing arithmetic in Roman numerals is probably the prime example of complification by inhomogeneity of the rules —inhomogeneity that is invoked by inadequate notation— .

The variety of notations for quantification, however, is a still existing source of inhomogeneity. That is, no doubt, the reason why the corresponding rules of manipulation, which we consider teachable to freshmen in mathematics, do not always get the attention they deserve.

Yet, employing a uniform notation, with "$\underline{op}$" denoting a binary, symmetric, and associative operator, and writing "$\underline{OP}$" for its associated quantifier version, we, for instance, have

$$(\underline{OP}\, x : x = y : t.x) = t.y \quad ,$$

i.e. such operators all satisfy the same one-point rule, and for any idempotent $\underline{op}$ we have

$$(\underline{OP}\, x : P.x \vee Q.x : t.x) = (\underline{OP}\, x : P.x : t.x)\ \underline{op}\ (\underline{OP}\, x : Q.x : t.x)$$

(and similarly so for an arbitrary $\underline{op}$ and P and Q satisfying $P.x \not\equiv Q.x$ for all x), and we have

$$(\underline{OP}\, x : \textbf{false} : t.x) = \text{unit element of } \underline{op}\quad,$$

if a unit element exists.

Thus, with a uniform notation, the rules for manipulating universal and existential quantifications, for summations, products, etc. can be presented (and remembered!) all at once in a concise way. The rules mentioned above are not intended to be a complete summary. More extensive treatments can, for instance, be found in "Een methode van programmeren" by Edsger W. Dijkstra and W.H.J. Feijen —also in German and English— and in "Program Construction and Verification" by Roland C. Backhouse. Bird and Wadler's monograph presents another uniform notation for quantifications.

Yet another concept for which notational conventions vary widely is function application. Besides the juxtaposition of a function identifier and an argument within parentheses, subscription is often used as function application as well: an infinite sequence A_0, A_1, A_2, $\ldots$ of naturals is a natural function on the natural numbers. We might, therefore, write $A.i$, and $A.(i+j)$, and $A.(A.i)$ just as well. Note, however, that subscription is usually considered right-associative, while function application is left-associative.

There are quite a few other conventions for function application around, such as overscripts like in $\hat{P}$, $\bar{P}$, $\check{P}$, and $\overset{\infty}{P}$, or superscripts such as in P^*, P', and P^T, or bracket pairs such as in the notation $|P|$ for absolute value and number-of-elements, with P denoting the argument and $\hat{\ }$, $\bar{\ }$, $\check{\ }$, etc. denoting the function.

Upon closer scrutiny, these conventions present all kinds of problem.

• First of all there are the typographical problems, of symbol variety, but particularly of how to render the application of an overscript operator to an expression: what about $\overbrace{a+b*c}$ or $\overbrace{P \cap (Q \cup R)}$?

• More seriously, symbols like the prime sometimes do and sometimes do not denote a function: in, for instance, geometrical arguments the prime is often used to group identifiers — A, A', A'' versus B, B', B'', while A' and A'' do not necessarily have the same relation as A and A'; in other words, A'' does not stand for $(A')'$ —. It is not always simple to infer from the context which convention is meant.

• Function application usually has the greatest binding power, but how do we read an expression like $\hat{P}'$ if "^" and " $'$ " do not commute?

End • .

Again, as a consequence of inhomogeneity there are, first of all, the dangers of misinterpretation, and, secondly, the multitude of rules, which by their very number, are usually left implicit.

End Homogeneity .

Symmetry
Symmetry in the rules and the notation is another means of avoiding too much pluriformity. We discuss two examples.

It is not uncommon to see conjunction being given a higher binding power than disjunction. We can, for instance, see

$$P \wedge Q \vee Q \wedge R \vee R \wedge P$$

$$\equiv$$

$$(P \vee Q) \wedge (Q \vee R) \wedge (R \vee P) \quad ,$$

which has lost the nice symmetry of

$$(P \wedge Q) \vee (Q \wedge R) \vee (R \wedge P)$$

$$\equiv$$

$$(P \vee Q) \wedge (Q \vee R) \wedge (R \vee P) \quad .$$

Giving $\wedge$ and $\vee$ the same binding power reflects more explicitly that rules about $\wedge$ and $\vee$ usually come in pairs, coupled by De Morgan's Law.

In electrical engineering the symmetry has been obscured still further by the use of the symbol " $+$ " for disjunction and the invisible multiplication sign for conjunction, with the corresponding binding powers. Then the distributive rule $P(Q + R) = PQ + PR$ looks very familiar, but the companion rule $P + QR = (P + Q)(P + R)$, looking somewhat strange, is much less well known among electrical engineers. That is an unfortunate consequence of the notational choice.

The other example of destroying symmetry that we discuss is the notation for binomial coefficients. $\binom{n}{k}$ is a function of n, k, and $n - k$, and it is symmetric in k and $n - k$. The traditional notation, by exhibiting one of the latter two plus the sum n, obfuscates this symmetry completely; it now has to be cast in the form of a theorem: $\binom{n}{k} = \binom{n}{n-k}$. The pluriformity immediately extends to other formulae: the well-known relation $\binom{n}{k} = \binom{n-1}{k-1} + \binom{n-1}{k}$ is equivalent with $\binom{n}{k} = \binom{n-1}{k} + \binom{n-1}{n-k}$, which might even be considered the nicer form because it exhibits the symmetry between k and $n - k$.

Consider the alternative notation that exhibits the summands as parameters: for natural i and j, we write $P.i.j$ instead of $\binom{i+j}{i}$ and $\binom{i+j}{j}$. Now $P.i.j$ is only asymmetric in that we do not have a symmetric linear representation for an unordered pair of arguments. The asymmetry in the rule $\binom{n}{k} = \binom{n-1}{k} + \binom{n-1}{k-1}$ also disappears, and hence so does the dilemma which of its forms to choose: with some renaming it becomes

$$P.i.j \; = \; P.i.(j-1) + P.(i-1).j \quad .$$

In view of the latter, a nice recursive definition of $P.i.j$ is

$$
\begin{aligned}
P.i.j \; = \quad &\textbf{if } i = 0 \lor j = 0 \;\rightarrow\; 1 \\
&[\!] \; i \neq 0 \land j \neq 0 \;\rightarrow\; P.(i-1).j + P.i.(j-1) \\
&\textbf{fi} \quad .
\end{aligned}
$$

With the alternative notation we can now write

$$(a+b)^n = (\underline{S}\, i,j : i+j = n : P.i.j * a^i * b^j)$$

instead of the traditional asymmetric formula. Mathematics has a vast reservoir of identities concerning binomial coefficients. We wonder, however, how much the size of that reservoir may be cut down by using the alternative notation. Recently we, for instance, encountered in a summary of some such useful identities both the formula $(\sum k :: \binom{r}{k}\binom{s}{n+k}) = \binom{r+s}{r+n}$ and the formula $(\sum k :: \binom{r}{k}\binom{s}{n-k}) = \binom{r+s}{n}$ for natural r and integer n, while both could have been written much more symmetrically in the form

$$P.a.b \; = \; (\underline{S}\, i,j : i+j = r : P.i.j * P.(a-i).(b-j)) \quad ,$$

the one with $a,b := r+n, s-n$ and the other with $a,b := r+s-n, n$.

End Symmetry .

*

So much for what can be involved in bridling the size of the body of rules. Another source of inconvenience can be found in rules of manipulation that do not provide enough combinatorial freedom.

Combinatorial freedom

From the remote past, we have the example of De Morgan who, for quite some time, did not have a symbol for negation. Because negation is its own inverse, it pairs propositions; using the one-to-one correspondence between small and capital letters, De Morgan denoted the negation of Q by q. But the convention does not cover negation of expressions, so that to formulate, say, De Morgan's Law, one has to go through contortions like "if $R = p \vee q$ then $r = P \wedge Q$" —arbitrarily choosing one of the eight forms of the formula— .

There are similar phenomena in more recent literature, e.g. in R. Courant's "Differential and Integral Calculus": Courant writes "If $\phi(x) = f(x) + g(x)$, then $[\ldots]$ $\phi'(x) = f'(x) + g'(x)$" where the formula $(f + g)' = f' + g'$ would have expressed more directly that differentiation distributes over function addition. Courant's reluctance to introduce expressions whose values are functions, forces him to introduce the name ϕ, so as to have something to attach the prime to.

(Whether Courant's reluctance was a consequence of unfamiliarity with the concept or whether he had other reasons is not very relevant here. The point we want to make is that purely manipulative considerations may invite one to investigate such expressions, because in their wake it might be possible to formulate attractive manipulation rules like the "distribution" $(f + g).x = f.x + g.x$ and the earlier mentioned $(f + g)' = f' + g'$.)

Yet another type of lack of combinatorial freedom is what we might call "combination of two operators into one". Consider, for example, equality of naturals modulo some fixed N, denoted by $a \equiv b$ (mod N). There are special symbols $\oplus$ and $\otimes$ for addition modulo N and multiplication modulo N of two natural numbers. Can we express, with these notational means only, the reduction of a number n modulo N? We can, but we have to use $\oplus$ —$n \oplus 0$— or $\otimes$ —$n \otimes 1$— . It is much more convenient, however, to use a unary arithmetic operator reduction-modulo-N. Having it available makes the introduction of special symbols like $\oplus$ and $\otimes$ superfluous.

A similar situation arises in predicate logic. We have the boolean connective $\equiv$, equivalence. It is sometimes suggested that "$P \equiv Q$ is everywhere true" be denoted by $P = Q$, i.e. that a special symbol $=$ be used to denote the universal equivalence of two predicates. The convention raises the same problem as the one arising in the previous example: do we also need special symbols for universal implication, disjunction, difference, etc.? And if we think we do, do we like it that then $\neg(P = Q)$ and $P \neq Q$ are in general not the same? And how do we express "P holds everywhere": as "$P = \textbf{true}$"? In addition we have to ask ourselves where to write an "$=$" in $P \equiv Q \equiv R$ to denote that the predicate holds everywhere. Again, a unary "is-everywhere-true" operator is a way out. Dijkstra and Scholten, in EWD813, have suggested a special bracket pair "[" and "]", pronounced "everywhere", denoting universal quantification over some universe to be specified.

Remark . For functions f and g in general, $f = g$ is often used to denote universal equality of f and g. In analogy with $f + g$, however, we could also define it by $(f = g).x \equiv (f.x = g.x)$ for all x, i.e. as a boolean function on the same domain as functions f and g that is **true** whenever f and g are equal and **false** otherwise. In other words, the first convention can only be used in a context where there is no need to express this boolean function.
End Remark .

End Combinatorial freedom .

Finally, in this discussion of convenience and simplicity of manipulation, the boolean connectives $\equiv$ (equivalence) and $\Leftarrow$ (follows-from) deserve mention. The symbol $\Leftarrow$ is the notational instrument giving us the freedom to present calculations as a sequence of strengthenings rather than the traditional sequence of weakenings, if we so desire. As discussed more extensively elsewhere (in Chapter 14) this is not just a cosmetic affair.

We particularly mention the equivalence here, because from the point of view of manipulation it is very attractive: it obeys Leibniz's Rule of substitution of equals for equals, and such substitution is the simplest type of manipulation one can imagine.

In mathematical arguments, often many steps rendered as implications are equivalent reformulations; each definition, for instance, is an equivalence, and, hence, each substitution of the definition is an equivalent reformulation. It is not always advantageous or possible to stick to equivalences when massaging a demonstrandum, but we should realize that the option is there.

*   *
*

16.2   Avoiding formal laboriousness

From a formalism to be used with pen and paper, we require that it offer the possibility of avoiding lengthiness and duplication in formulae and proofs.   (Such a possibility for economy of expression is the least one may expect from a formalism; judicious use will, however, always be necessary as well.)   We shall discuss three things: what rules to include, how the proof format assists in avoiding laboriousness, and how notation and naming can do so as well

*   *

The fewer rules a formalism comprises, the longer the proofs carried out in it.   Hence, if brevity is deemed important —and, here, for us it is— the introduction of additional, derived rules is similarly important.   As for which properties to add as derived rules, the best exploitable properties are those that justify simple manipulations. They are valuable not only because simplicity of manipulation is what we are after, but also in the sense that a property that is difficult to formulate will be of less heuristic guidance.

There are many examples of nicely exploitable properties. Symmetric rewrite rules, i.e. rules of the form $P = Q$, or $[P \equiv Q]$ for predicates P and Q, are the prime example, because substitution of equals for equals is such a simple manipulation.

Particularly boolean equivalence has proved to be useful for keeping proofs short. It is our experience that, were we to disallow direct exploitation of equivalences in our predicate calculus —as is the case in, for instance, Gentzen's system of Natural Deduction proper— the length of proofs might increase by a factor 2 to 8 or more: if we have to use, or prove, $P \equiv Q$ by using, or proving, mutual implication $(P \Rightarrow Q) \wedge (Q \Rightarrow P)$, or $(P \wedge Q) \vee \neg(P \vee Q)$, or $(\neg P \vee Q) \wedge (P \vee \neg Q)$, rather than having the extra possibility of exploiting symmetric rewrite rules, the danger of doublings in length of the proof or width of the formulae is already built in. If a formula contains more than one equivalence sign, the differences in length may become more dramatic. As Roland C. Backhouse shows in "Program Construction and Verification", pp.49–51, in Gentzen's system the simplification of $s \equiv (s \not\equiv g)$ into $\neg g$ may take more than 20 steps (the precise number of steps depends on how the equivalences are rendered); this is in sharp contrast with the three steps needed in

$$s \equiv (s \not\equiv g)$$
$$= \qquad \{\text{definition of } `` \not\equiv \text{''}\}$$
$$s \equiv (s \equiv \neg g)$$
$$= \qquad \{\text{associativity of } `` \equiv \text{''}\}$$
$$(s \equiv s) \equiv \neg g$$
$$= \qquad \{p \equiv p \text{ is identity element of } `` \equiv \text{'' for any } p\}$$
$$\neg g \quad .$$

(The latter example is due to J.G. Wiltink, Information Processing Letters.)

Besides equivalences, other nicely exploitable properties are distributive properties, or more generally, homomorphisms, viz. operations

f satisfying $f.(a \,\square\, b) = f.a \,\diamond\, f.b$ for some operators $\square$ and $\diamond$; also monotonicity properties are useful.

As for other rules to include, brevity is served if we can avoid such low-level manipulations as the rearrangement of parentheses or the reshuffling of terms in a formula. So it helps if we are allowed to exploit the associativity of an operator by leaving out parentheses altogether, and to exploit symmetry (combined with associativity) by performing arbitrary permutations of terms in one go or, if the permutation is intended to prepare another transformation, not perform the permutation at all and appeal to the symmetry explicitly instead.

*   *

So much for the choice of derived rules. We now turn to the proof format. Traditionally, a proof consists of a sequence of statements, the validity of each following —by some explicitly stated or implicitly used deduction rule— from earlier statements. Intermediate statements play a double rôle: they act as conclusions from earlier statements in one step and as premisses in a later step. Despite their double rôle they are usually mentioned only once. Later use as a premiss is either left implicit or made explicit by means of labelling and referencing.

In the calculational proof format that we use, duplication is avoided by notational means in two ways. Firstly, this format too has been chosen so as to avoid duplication of intermediate expressions: a proof of $[P \equiv Q]$ that consists in transforming P into R, i.e. proving $[P \equiv R]$, and subsequently transforming R into Q, i.e. proving $[R \equiv Q]$, is rendered as

$$
\begin{array}{ll}
P & \\
= & \{\text{hint why } [P \equiv R]\} \\
R & \\
= & \{\text{hint why } [R \equiv Q]\} \\
Q & ,
\end{array}
$$

so that duplication of R is avoided.

Secondly, a proof of $[P \equiv Q]$ may take the forms

$$
\begin{array}{ccc}
\begin{aligned}
& P \\
=& \\
& \vdots \\
=& \\
& P \wedge Q \\
=& \\
& \vdots \\
=& \\
& Q
\end{aligned}
&
\text{or}
&
\begin{aligned}
& P \\
=& \\
& \vdots \\
=& \\
& P \vee Q \\
=& \\
& \vdots \\
=& \\
& Q \quad .
\end{aligned}
\end{array}
$$

In such calculations, first P has to be dragged along while Q is introduced, and subsequently Q has to be dragged along while P is removed again. The repetition of expressions P and Q is quite cumbersome.

Such repetitions can be avoided because we allow steps that are not equivalences: $P \Rightarrow Q$ (and the equivalent $Q \Leftarrow P$) standing for $P \wedge Q \equiv P$ and $Q \vee P \equiv Q$ enables us to render a calculation establishing $[P \wedge Q \equiv P]$ by

$$
\begin{aligned}
& P \\
\Rightarrow & \quad \{ * : \text{hint why } [P \Rightarrow R] \} \\
& R \\
\Rightarrow & \quad \{ ** : \text{hint why } [R \Rightarrow S] \} \\
& S \\
\Rightarrow & \quad \{ \text{hint why } [S \Rightarrow Q] \} \\
& Q \quad ,
\end{aligned}
$$

which in the absence of implication would have to be rendered as

$$
\begin{array}{ll}
 & P \\
= & \{*\} \\
 & P \wedge R \\
= & \{**\} \\
 & P \wedge R \wedge S \\
= & \{\text{hint why } [S \wedge Q \equiv S]\} \\
 & P \wedge R \wedge S \wedge Q \\
= & \{**\} \\
 & P \wedge R \wedge Q \\
= & \{*\} \\
 & P \wedge Q \quad ,
\end{array}
$$

a calculation that is both longer and wider.

For demonstranda of the form $[P \Rightarrow Q]$ there is yet another calculational format that we use, viz. a calculation that reduces Q to **true** in the scope of P, meaning that P is allowed to occur in hints. That format is sometimes more convenient, for instance if P is used only once in the middle of a calculation.

So much for notational and other conventions in the proof format that help avoid laboriousness. In addition, judicious use of the format is needed. For demonstranda of the forms $[P \equiv Q]$ and $[P \Rightarrow Q]$, we always have the choice between a calculation manipulating the whole expression $P \equiv Q$ and $P \Rightarrow Q$, reducing it to **true**, and a calculation that transforms one side into the other. The second type is more restrictive: it precludes manipulations involving both sides at the same time, like $P \Rightarrow Q$ being rewritten into $\neg P \vee Q$. It is, however, the better option if all manipulations involve one side only: in such a situation, manipulating the whole expression would mean that in each step the unmanipulated side would have to be copied. Even if this copying were avoided by combining independent manipulations of the sides in one step, the gain would be minor: what is gained in brevity of the calculation is lost in width (of the formulae), and although there would be fewer hints, they would be twice as long and, as argued in section 14.0 on the division of labour, necessarily more specific.

In our experience, the calculational format has a few more nice properties.

- Firstly, it enables us to distinguish in a very concise way between steps that are equivalences and steps that are strengthening or weakening. That possibility is particularly useful if one is designing a proof: massaging a demonstrandum or premiss into an equivalent one is not a commitment yet, while strengthening (a demonstrandum) or weakening (a premiss) is a design decision that may preclude completion of the proof.

Remark . Having the notational means of making equivalences explicit has made us much more aware of the presence of such equivalences in arguments. As we alluded to earlier, there are many more than we had realized before.
End Remark .

- Secondly, unlike in the "traditional" proof format, a step in the calculation is a very local affair: involved are the first expression, the hint, and the second expression, and nothing else; the validity of that step does not depend on the prefix of the calculation. That is, the structure of such a calculational proof is truly linear.

As we mentioned earlier it is not impossible to "reuse" a formula later on in the calculation, but then it has to be dragged along through the calculation and added as a conjunct or a disjunct to all intermediate expressions: if we want to reuse P we might write $P \Rightarrow Q \wedge P$ instead of $P \Rightarrow Q$, $P \equiv Q \wedge P$ instead of $P \equiv Q$, and $P \Leftarrow Q \vee P$ instead of $P \Leftarrow Q$. Such recurrence of mostly unexploited parts of formulae, however, is something we would rather avoid. We tend to consider it an indication that some recasting of the argument might be due: we might be forcing an argument into a linear shape that is not suited for it.

The locality of steps in a calculation is certainly one of the advantages of the linear format. Yet we do sometimes find it useful to deviate from it. The following calculation is a typical illustration.

Example .   Given $[(\underline{A}z :: X)] \equiv [X]$ for all X , we prove $[(\underline{A}z :: X) \vee Y] \equiv [(\underline{A}z :: Y) \vee X]$, for any X and Y :

$$[(\underline{A}z :: X) \vee Y]$$

= $\qquad$ {premiss with $X := (\underline{A}z :: X) \vee Y$ }

$$[(\underline{A}z :: (\underline{A}z :: X) \vee Y)]$$

= $\qquad$ { $\vee$ distributes over $\underline{A}$: first disjunct is independent of z }

$$[(\underline{A}z :: X) \vee (\underline{A}z :: Y)]$$

= $\qquad$ {undoing with $X, Y := Y, X$, using symmetry}

$$[(\underline{A}z :: Y) \vee X] \quad .$$

End Example .

In the example, the last hint "undoing" indicates a reusing of the calculation so far, here instantiated by $X, Y := Y, X$, to justify the last step. We sometimes prefer this rendering of the argument to its alternative, viz. isolating the reused relation as an explicitly formulated lemma, if that relation is not significant elsewhere and if the calculation establishing it is of the type "there is hardly anything else you can do".

• $\qquad$ A third advantage of the calculational style is that the validity of the steps is independent of the truth value of the expressions massaged: a step $P = \{\ldots\} Q$ holds irrespective of the value of, for instance, P . As a result, proof by contradiction loses much of its special status: a calculation of the form $\neg P \Rightarrow \{\} \ldots \{\}$ **false** is not more special than a calculation of the form $P \Leftarrow \{\} \ldots \{\}$ **true** . In fact, the one can be transformed into the other, by transforming steps $R \Rightarrow Q$ into $\neg R \Leftarrow \neg Q$, and $R \equiv Q$ into $\neg R \equiv \neg Q$. We, for instance, have an infinite number of primes exists

$=$ {definition of "infinite number"}

for every finite set of primes there is a prime not in the set

$\Leftarrow$ { }

for every finite set V of primes: some prime divides $(\prod p : p \in V : p) + 1$ and no prime in V does so

$\Leftarrow$ {factorization and divisibility properties}

no prime divides 1

$=$ { }

true .

In other words, a proof by contradiction that has such a linear form can be avoided.

End $\bullet$.

So much for a discussion of how the calculational format contributes to the avoidance of laboriousness, and of some more of its properties.

$$* \qquad *$$
$$*$$

Notation plays a major rôle in the avoidance of lengthiness and duplication. We discuss a number of common examples.

• The earlier mentioned coexistence of implication and equivalence introduces redundancy: $X \Rightarrow Y$ and $X \wedge Y \equiv X$ or $X \vee Y \equiv Y$ are semantically equivalent, and so are $X \equiv Y$ and $(X \Rightarrow Y) \wedge (Y \Rightarrow X)$; since expressing the one into the other in both cases means duplication of at least one argument, we want to keep both so as not to be forced to use the one where the other would give a shorter formula or proof. For similar reasons, mathematics uses both $x \leq y$ and the equivalent $x \underline{\max} y = y$.

- From a manipulative point of view, the ...-notation for quantified expressions is unfortunate. It forces one to give at least two or three terms of the quantified expression so as to avoid ambiguities — $L_1 x + \frac{L_2}{2} x^2 + \cdots + \frac{L_n}{n} x^n + \cdots$ or $1 + xy^2 + x^2 y^4 + \cdots$ — ; that means that any manipulation that massages the term of such an expression has to be done in morefold. And how about rendering a range split, i.e. a manipulation like $(\underline{S} i : 0 \leq i < n : t.i) = (\underline{S} i : 0 \leq i < k : t.i) + (\underline{S} i : k \leq i < n : t.i)$, in the ...-notation? Other shortcomings can be seen in an expression like $\overbrace{a_1 + a_1 + \cdots + a_1}^{n_1}$. (Yet another problem with the notation is whether it is supposed to denote empty ranges as well.)

- If $x_1, \ldots, x_k$ were our only notation for finite sequences, we would always be forced to introduce names for the length of the sequence and for each of its elements. We would, for instance, have to write $\mathrm{wp.}(\text{``}x_1, \ldots, x_k := E_1, \ldots, E_k\text{''}, R) = R_{E_1, \ldots, E_k}^{x_1, \ldots, x_k}$. Naming the sequences rather than their elements, we can write $\mathrm{wp.}(x := E, R)$ and R_E^x.

Not only does such frugality with nomenclature keep the expressions shorter, it can also be essential for developing a calculus on some domain. Consider, for the sake of the discussion, permutations on some finite universe. Expression $(a_0 a_1 \ldots a_n)$ usually denotes a permutation that maps a_i onto a_{i+1} —addition to be taken modulo $n + 1$— for $0 \leq i \leq n$ and leaves all other elements of the universe unchanged. We might consider functional compositions of such (cyclic) permutations and try to formulate some calculus for them. Unfortunately, however, even a simple rule like $(a_0 a_1 \ldots a_n) = (a_k a_{k+1} \ldots a_n a_0 a_1 \ldots a_{k-1})$ is cumbersome to formulate and hence to use.

If we introduce the notation $[A]$, for instance, for sequence A of distinct elements only, instead of the above $(a_0 a_1 \ldots a_n)$, we can render the rule mentioned above by

0. $[AB] = [BA]$.

Similarly, all kinds of other rule can now be written down economically. For instance, with X and Y sequences, p and q elements, and each expression between "[" and "]" denoting a sequence of distinct elements:

1.　　　$[X] \circ [Y] = [Y] \circ [X]$ for disjoint X and Y ;

2.　　　$[XpY] = [Xp] \circ [pY]$;

3.　　　$[p] \circ [X] = [X]$, $[X] \circ [p] = [X]$, i.e. $[p]$ is the identity element of $\circ$;

4.　　　$[pq] \circ [pq] =$ identity element of $\circ$; etc. .

Using these rules we can then calculate with permutations, as in the following proof of $[XpYq] \circ [pq] = [Xp] \circ [Yq]$:

$$[XpYq] \circ [pq]$$
$$= \qquad \{\, 2. \text{ with } Y := Yq \,\}$$
$$[Xp] \circ [pYq] \circ [pq]$$
$$= \qquad \{\, 0. \text{ on middle term} \}$$
$$[Xp] \circ [Yqp] \circ [pq]$$
$$= \qquad \{\, 2. \text{ on middle term with } X, p, Y := Y, q, p \,\}$$
$$[Xp] \circ [Yq] \circ [qp] \circ [pq]$$
$$= \qquad \{\, 4., \text{ using } [qp] = [pq] \text{ on account of } 0. \}$$
$$[Xp] \circ [Yq] \quad .$$

The formulation of the above rules and their use in the calculation shown, would, we think, have been almost impossible with the original notation. (Indeed, we have never seen them stated in the literature. They arose in the exploration of some programming problems dealing with permutations. (See Chapter 12.))

For us, these explorations with permutations once more confirmed that the introduction of subscripted variables requires great care and frugality. All too soon, there are so many subscripts around that manipulation is truly hampered. It appears that if one only introduces subscripts if they are hard to avoid, many of the ones that can be seen in the literature are avoidable.

A particular source of subscripts, creeping into correctness proofs of programs, is the use of arrays. (Our first treatments of the earlier mentioned permutation problems were full of subscripts and as a result hardly convincing.) Often, the array is best considered an implementation device, in which one expresses the notions that are most helpful in the design stage as a separate activity.

•   Finally we note that naming conventions play a rôle in the avoidance of lengthiness and duplication. We mentioned subscripted variables already in the above. The topic is discussed more extensively in Chapter 15.

End • .

With this discussion of formal laboriousness we end this chapter on the use of formalism.

17 Epilogue

"If in its continual development mathematics seldom if ever attains a finality, the constant growth does mature some residue that persists. But it is idle to pretend that what was good enough for our fathers in mathematics is good enough for us, or to insist that what satisfies our generation must satisfy the next."

Eric Temple Bell
"The Development of Mathematics"
McGraw-Hill Book Company
New York London 1945, p.172

The initial incentive to this study was above all the need to improve on the status quo and develop a style of reasoning in which clarity and convincingness of exposition go together with the detail and precision required from the correctness proofs of programs.

Initially, we were primarily concerned with form and presentation, because in their tangibility they were things we could experiment with. Findings of these experiments are dealt with in Chapter 14 [Clarity of exposition]. Experiments with fine-grained reasoning were a main concern, and a particularly important question, then, was how to organize a detailed argument so as to keep it manageable.

Top-down presentation of proofs is one answer to this question, because it helps defer detail, in verbal and in formal reasoning alike. The resort to formalism is another answer; a great deal of experimentation was devoted to it. The calculational style of reasoning was stressed, again for the sake of organization and manageability: not only does it make the structure of an argument more visible, the calculational format lends itself more readily to detailedness and conciseness at the same time.

A result of these experiments, which were performed in cooperation with Edsger W. Dijkstra and W.H.J. Feijen and which were joined by others in their teaching and writing, is that over the years it has become possible to present all kinds of ingenious algorithms, in all their necessary detail, in much less space than before. To name just a few: Shiloach's algorithm for checking the equivalence of two circular lists [Chapter 13], Boyer and Moore's Majority Vote algorithm, Heapsort, distributed termination detection algorithms, and the like.

As time went by we experienced, through our concerns for form and presentation, that deeper issues were involved. For instance, viewed as reference to what they stand for, names can be regarded as mere presentational devices for shortening the text; but they appeared to emerge as carriers for abstraction and decomposition, for confinement of detail, and postponement of commitment during the development of a design. Consequently, our concerns extended themselves to streamlining the arguments themselves.

Along the way, the traditional distinction between form and content began to fade. The experience that the introduction of nomenclature could be a design decision, influencing the shape of the resulting argument, was one of the circumstances causing the distinction to fade. There were others; for instance, the introduction of $A \Leftarrow B$ as a notational alternative to $B \Rightarrow A$ turned out to be much more than that: it provided richer linguistic means for expressing derivations in which steps are of the type "there is hardly anything else you can do". The richer linguistic means eventually provided us with more heuristic guidance in the design process, thus gradually bringing us into the realm of heuristics.

As mentioned already in the introduction to this study, this is only the beginning of a larger investigation. There is, first of all, the confinement to the task of proving one theorem or designing one program. Building up a theory is a game for which the rules of thumb developed here will undoubtedly be insufficient and partly inappropriate. (It is, for instance, hard to see how or why to develop a theory in top-down fashion unless the theory is meant to solve one particular problem.)

This is also a beginning in the sense that we have chosen for breadth rather than depth in the explorations. Our strategy in selecting material for experimentation has largely been to pick out problems and proofs that according to our standards of the day were not dealt with satisfactorily, and to try to find and remedy the trouble. The remedies were as varied as the problems, and that is how we encountered topics like fine-grainedness of detail as well as organization and arrangement; the choice of notation as well as issues of naming; avoiding case distinctions and proofs of equivalence conducted by proof of mutual implication as well as the heuristic guidance offered by the shape of a formula; the avoidance of proof by contradiction as well as the exploitation of a calculational style.

In many of the single subjects, however, a lot remains to be done. Take case analysis, as an example. On the one hand, as most mathematicians will agree, it is often best avoided, and we have seen that the choice of nomenclature may play a rôle in this, because it may destroy symmetries or introduce overspecificity. On the other hand, we are still far removed from having extensive criteria for deciding when a case distinction is avoidable and when it is not. Even though it is idle to hope for the rule without exception, a systematic exploration could be very helpful.

The situation with equivalences proved by proving mutual implication is similar. Admitting equivalence as a full-blown connective has made us more aware of cases in which mutual implication can be avoided —avoidance usually having the effect of shortening the proof— . There are also situations where avoiding mutual implication is clearly not possible. (Consider, for instance, demonstrandum $[P \equiv Q]$ where one of the premisses to be exploited has the shape "for all X, $[f.X \Rightarrow X] \Rightarrow [P \Rightarrow X]$".) Again, the topic deserves a much more extensive dedicated exploration. Case analysis and proof by mutual implication are by no means the only interesting objects of further exploration.

In a way, we are constantly trying to hit a moving target. At a time when we were confining our attention to form and presentation, an objection put forward was that proving rather than presenting or streamlining theorems was the difficult task, and later when heuristic considerations had entered the scene and we had begun concentrating on "given a theorem, design a proof" as a task, a new objection was that finding a theorem might be more difficult than proving it. There is truth, no doubt, in this ranking of difficulties, and, holding the view that there is no point in trying to master the difficult before you know how to deal with the simpler, we explored presentation first and heuristics only much later. As for finding theorems: just as the boundary between form and content has proved to be less sharp than is usually considered, in the same way the distinction between proving and finding theorems is somewhat diffuse: we are continually formulating additional theorems in the course of a proof, even though these may not be of interest in themselves.

Finally, in the methodological explorations discussed here, our main intention has been to be as explicit as possible about criteria for exposition and design that we have come to value and profit from. In part, what we did may perhaps be considered as making explicit styles of reasoning that able mathematicians use all the time —more or less subconsciously— . Some may even hold the view that it is not more than that, just like in computing many held, and perhaps still hold, the opinion that the programming methodology developed in the seventies was not necessary for the good programmer because he already worked along those lines subconsciously.

Even if it is all we have done, we believe we have gained a lot, because explicitness has two, not unrelated, advantages; the development of computing into a teachable scientific discipline exemplifies both these advantages. First of all, explicitness moves the boundary between what is considered simple and what counts as difficult; it "increases the number of operations that can be performed without thinking about them" —we quote Whitehead— .

Secondly, in the wake of the aforementioned advantage, such explicitness paves the way to teachability, not only of problems that were too difficult or advanced before, but also of the methodological issues involved. In programming, experiences have become more and more favourable in the last decade. In mathematics at large, such experiences in teaching methodological issues explicitly are much rarer. It is our strong belief that the teaching of such a topic is both possible and profitable.

18 Proof rules for guarded-command programs

For predicates P and Q and program or program fragment S, $\{P\}S\{Q\}$ is a boolean expression, whose operational interpretation is that execution of S when started in a state satisfying P terminates in a state satisfying Q. P is called the precondition of S, and Q is its postcondition. (We note that in contrast with, for instance, Hoare triples, here triple $\{P\}S\{Q\}$ denotes total correctness, i.e. includes termination.)

Solving a programming problem often means solving an equation $(S : \{P\}S\{Q\})$. The following rules are convenient for calculating with such equations.

$$\{P\}\, S\, \{Q\} \qquad \Leftarrow \{P\}\, S\, \{R\} \,\wedge\, [R \Rightarrow Q]$$
$$\{P\}\, S\, \{Q\} \qquad \Leftarrow [P \Rightarrow R] \quad \wedge\, \{R\}\, S\, \{Q\}$$
$$\{P\}\, S\, \{Q \wedge R\} \equiv \{P\}\, S\, \{Q\} \,\wedge\, \{P\}\, S\, \{R\}$$

We define the constructs of the program notation in terms of expressions $\{P\}S\{Q\}$:

- skip : $\{P\}\, \text{skip}\, \{Q\} \equiv [P \Rightarrow Q]$

- assignment : $\{P\}\, x := E\, \{Q\} \equiv [P \Rightarrow Q(x := E)]$

- composition : $\{P\}\, S; T\, \{Q\} \equiv$ there exists an H such that
  $$\{P\}\, S\, \{H\} \,\wedge\, \{H\}\, T\, \{Q\}$$

171

- conditional statement :

$$\{P\} \text{ if } B.0 \rightarrow S.0 \: [] \: \ldots \: [] \: B.n \rightarrow S.n \text{ fi } \{Q\}$$

$$\equiv$$

$$(\underline{A} i : 0 \leq i \leq n : \{P \wedge B.i\} \: S.i \: \{Q\})$$

$$\wedge$$

$$[P \Rightarrow (\underline{E} i : 0 \leq i \leq n : B.i)]$$

- repetitive statement :

We deal with repetitions of the shape $\text{ do } B \rightarrow S \text{ od }$ only, because
$\text{do } B.0 \rightarrow S.0 \: [] \: \ldots \: [] \: B.n \rightarrow S.n \text{ od }$ is considered to be the same program as
$\text{do } (\underline{E} i : 0 \leq i \leq n : B.i) \rightarrow \text{ if } B.0 \rightarrow S.0 \: [] \: \ldots \: [] \: B.n \rightarrow S.n \text{ fi } \text{ od }$.

A proof of the validity of $\{P\} \text{ do } B \rightarrow S \text{ od } \{Q\}$ as a rule consists of three parts: for some predicate H

- a proof that it satisfies $\{H\} \text{ do } B \rightarrow S \text{ od } \{H \wedge \neg B\}$

- a proof of $[P \Rightarrow H]$

- a proof of $[H \wedge \neg B \Rightarrow Q]$.

The first part is proved by an appeal to the Invariance Theorem, which states that

Theorem . $\{H\} \text{ do } B \rightarrow S \text{ od } \{H \wedge \neg B\}$ follows from

- H is invariant, i.e. $\{H \wedge B\} \: S \: \{H\}$ holds,
- termination is guaranteed, i.e. for some expression t in terms of the variables of the state space —the "variant function"— and a well-founded set $(C, <)$,
  $[H \wedge B \Rightarrow t \in C] \wedge \{H \wedge B \wedge t = X\} \: S \: \{t < X\}$ holds.

End Theorem .

Note . Frequently, C is chosen to be a subset of the integers that is bounded from above or below.
End Note .

End • .

Finally we mention that, viewed as an equation in P, $\{P\}\,S\,\{Q\}$ has a unique weakest solution, the weakest precondition, denoted as wp.(S,Q) . Then we have

$$\{P\}\,S\,\{Q\} \equiv [P \Rightarrow \text{wp.}(S,Q)] \quad .$$

In terms of wp , we can state, for instance, $[\text{wp.}(\text{skip}, Q) \equiv Q]$, and $[\text{wp.}(x := E, Q) \equiv Q(x := E)]$, and $[\text{wp.}(S; T, Q) \equiv \text{wp.}(S, \text{wp.}(T, Q))]$, etc. . The use of weakest preconditions often smoothens the derivation of a program.

19 Notational conventions

Logical operators .

$\neg$, $\wedge$, $\vee$, $\Rightarrow$ denote negation, conjunction, disjunction, and implication respectively as usual;

[ ]   pronounced "everywhere" or "for all states", is a bracket pair denoting universal quantification over a universe to be specified;

$\equiv$ pronounced "equivales" or "is equivalent to", is the symmetric associative connective usually denoted by $\Leftrightarrow$ or "if and only if";

$\not\equiv$ pronounced "differs from", is defined by $[(p \not\equiv q) \equiv \neg(p \equiv q)]$ for all p , q ;

$\Leftarrow$ pronounced "follows from", is defined by $[(p \Leftarrow q) \equiv (q \Rightarrow p)]$ for all p , q .

In decreasing binding power we have $\neg$; $\wedge$ and $\vee$; $\Rightarrow$ and $\Leftarrow$; $\equiv$ and $\not\equiv$.

The proof format .
Many of our proofs have the shape

$$
\begin{array}{ll}
P & \\
= & \{\text{hint why } [P \equiv Q]\} \\
Q & \\
\Rightarrow & \{\text{hint why } [Q \Rightarrow R]\} \\
R & , \\
\vdots &
\end{array}
$$

as a shorthand for $[P \equiv Q] \wedge [Q \Rightarrow R] \wedge \ldots$. The format is discussed more extensively in section 16.2.

Quantified expressions .

A   stands for the more familiar $\forall$,

E   for $\exists$,

S   for $\sum$,

P   for $\prod$,

MAX , MIN for max and min respectively.

[ ]   denoting universal quantification, it shares its properties with A .

Its major use is in Leibniz's Rule: if predicates P and Q satisfy $[P \equiv Q]$, occurrences of P in an expression may be replaced by Q without affecting the value of the expression.

The general pattern for quantified expressions is

$$(\underline{Q}\, x : p.x : t.x) \quad ,$$

with $\underline{Q}$ a quantifier, x a list of dummies, $p.x$ a boolean expression in terms of the dummies —the range— , and $t.x$ the term of the quantification. (See also section 16.1.) The same or a similar pattern shows in the notation for sets, equations, and numeral quantification:

$$\{x : p.x : t.x\} \text{ is the set usually denoted by } \{t.x | p.x\}$$
$$(x : p.x) \text{ denotes equation } p.x \text{ with unknown } x$$
$$(\underline{N}\, x : p.x : t.x) = (\underline{S}\, x : p.x \wedge t.x : 1) \quad .$$

Relational and arithmetic operators .

Relational operators have higher binding power than boolean operators. Associative operators such as min (minimum), max (maximum), gcd (greatest common divisor), etc. are used in infix notation.

Functions .

o   denotes function composition;

.   denotes function application, is a left-associative infix operator and has highest binding power. Expression $f.x.y$ usually denotes the application of function f of two arguments to the ordered pair x, y ;

if so desired, it can also be viewed as the application of (higher-order) function $f.x$ to argument y. If we want to stress that the arguments form a *pair* —or, in general, a vector— we write $f.(x, y)$.

Program notation .
See Chapter 18, for "$\{P\}S\{Q\}$", "skip", "$x := E$", "$;$", "**if** ... **fi**", "**do** ... **od**", and "wp". The bracket pair "$|[,]|$" opens and closes the scope of the variables declared after "$|[$".

Substitution .
$P(x, y := E0, E1)$: denotes expression P with all free occurrences of variables x and y replaced by expressions $E0$ and $E1$ respectively. The result is as if $E0$ and $E1$ are first evaluated and then substituted simultaneously. The notational convention is also used to express instantiations, as in "Theorem0 with $x, y := E0, E1$".

Set operators .
Operators $\cap$, $\cup$, $\subseteq$, $\supseteq$, $\subset$, $\supset$, $\div$, and $\in$ have their usual meaning; set complement is denoted by an infix "$\backslash$" or, where we discuss and quote Apostol, by "$-$"; "ϕ" denotes the empty set, and braces "$\{$" and "$\}$" are used for enumerated sets.

References

Apostol, Tom M.
Mathematical Analysis A Modern Approach to Advanced Calculus
Addison-Wesley Publishing Company, Reading Massachusetts, 1957.

Arbib, Michael A., Kfoury, A.J., and Moll, Robert N.
A Basis for Theoretical Computer Science
Springer-Verlag, New York, 1981.

Backhouse, Roland C.
Program Construction and Verification
Prentice-Hall International (UK) Ltd, London, 1986.

Bell, Eric Temple
The Development of Mathematics
McGraw-Hill Book Company, New York, 1945.

Bird, Richard and Wadler, Philip
Introduction to Functional Programming
Prentice-Hall International (UK) Ltd, London, 1988.

Brands, J.J.A.M. and Hautus, M.L.J.
Asymptotic Properties of Matrix Differential Operators
Journal of Mathematical Analysis and Applications 87, 1982, pp.199–218.

Courant, R.
Differential and Integral Calculus, Vol.I
Interscience Publishers, Second edition, 1937 (reprinted 1970).

Courant, Richard and Robbins, Herbert
What is Mathematics? An Elementary Approach to Ideas and Methods
Oxford University Press, London, 1956.

Dijkstra, Edsger W.
Selected Writings on Computing: A Personal Perspective
Springer-Verlag, New York, 1982.

Dijkstra, Edsger W. and Scholten, C.S.
About predicate transformers in general
EWD813, 1982.

Dijkstra, Edsger W. and Feijen, W.H.J.
Een methode van programmeren
Academic Service, The Hague, 1984.
(German (1985) and English (1988) translations by Addison-Wesley).

Dijkstra, Edsger W. and Gasteren, A.J.M. van
A monotonicity argument
Technical report AvG36/EWD878,
Eindhoven University of Technology, 1984.

Dijkstra, Edsger W. and Gasteren, A.J.M. van
A Simple Fixpoint Argument Without the Restriction to Continuity
Acta Informatica 23, 1986, pp.1–7.

Feijen, W.H.J., Gasteren, A.J.M. van, and Gries, David
In-situ inversion of a cyclic permutation
Information Processing Letters 24, 1987, pp.11–14.

Gasteren, A.J.M. van, and Feijen, W.H.J.
Shiloach's algorithm, taken as an exercise in presenting programs
Nieuw Archief voor Wiskunde XXX, 1982, pp.277–282;
issued by the Wiskundig Genootschap, Amsterdam.

Graham, Ronald L., Rothschild, Bruce L., and Spencer, Joel H.
Ramsey Theory
John Wiley & Sons, New York, 1980.

Halmos, P.R.
How to write mathematics
in P.R. Halmos, Selecta Expository Writing,
    ed. by Donald E. Sarason and Leonard Gillman,
    Springer-Verlag, New York, 1983.

Hardy, G.H., Littlewood, J.E., and Pólya, G.
Inequalities
Cambridge University Press, London, 1952.

Hardy, G.H. and Wright, E.M.
An Introduction to the Theory of Numbers
Oxford at the Clarendon Press, Third edition, 1954.

Huang, B.-C.
An algorithm for inverting a permutation
Information Processing Letters 12, 1981, pp.237–238.

Shiloach, Yossi
A fast equivalence-checking algorithm for circular lists
Information Processing Letters 8, 1979, pp.236–238.

Stark, Harold M.
An Introduction to Number Theory
Markham, 1970.

Valentine, Frederick A.
Convex Sets
McGraw-Hill Book Company, New York, 1964.

Whitehead, Alfred North
An Introduction to Mathematics
Home University Library — Oxford University Press, 1948.

Wiltink, J.G.
A deficiency of natural deduction
Information Processing Letters 25, 1987, pp.233–234.